CW00685407

As this is based on a true story, several names have been changed to maintain the privacy of a number of individuals.

My grateful thanks go to my parents, Matt and Peggy Walker, for their invaluable editorial skills; to Keith and Helen for help with the photography; to Jude, my sister-in-law, whose idea it was in the first place and - leaving the best until last - to Gordon, my lover and best friend, without whose help and encouragement this book would never have been completed.

British Library Cataloguing In Publication Data
A Record of this Publication is available
from the British Library

ISBN 1846853338
978-1-84685-333-3

First Published 2006 by

Exposure Publishing, an imprint of Diggory Press,
Three Rivers, Minions, Liskeard, Cornwall, PL14 5LE, UK
WWW.DIGGORYPRESS.COM

CHAPTER ONE
IN THE BEGINNING

The phone suddenly rang, and I snatched it up. I'd been sitting there waiting for it to ring for what seemed like an eternity.

'Hello?'

'Hello, it's Mom, I—.'

'Oh, Mom, it's you,' I interrupted. 'Sorry to cut you off, but I'm expecting Gordon to ring at any minute to tell me how he got on at the auction.'

'Of course! I'm so sorry. It completely slipped my mind. Will you let me know what happens?' she said.

'Yes, yes, don't worry. As soon as I hear anything I'll ring you. Bye.' I put the phone down quickly and expected it to ring again immediately. But no - total silence. This waiting around was beginning to seriously bother me.

Long after the auction should have been over I was still waiting for the expected call from Gordon. I was getting more and more agitated. Why didn't the dratted man ring me? All sort of scenarios were going through my mind, most of them gloomy and depressing. I needed to know – badly.

The phone eventually rang again and this time, to my relief, it was Gordon. There was a blast of noise at the other end, which could only mean that he was telephoning from a pub and, more than likely, had been there for some time.

'M, M, can you hear me?'

'Yes, yes, what's happened?' I gabbled.

'We've got it!' he yelled, 'we've bloody well got it!'

'Oh...oh...you clever chap!'

'I'm feeling pretty dammed pleased with myself.'

'How much?' was my next anxious question.

'Thirty-two thou! I just couldn't believe it. Still can't! It's an absolute miracle. I've met a few old friends and I'm just having a drink with them to celebrate,' he said ecstatically.

Thirty-two thousand pounds? When we were expecting to pay double, or even treble that amount was definitely cause for celebration. 'Wow, I can't believe it either. Well done, you.'

I felt drained with nervous tension, relieved and fit to celebrate myself. With only the dogs to join me, I decided to leave the bottle of champagne in the fridge until Gordon returned home and we could make merry in style. For us a new chapter was about to begin and we were both all set to accept the challenge.

Gordon and I made our living by renovating old houses. We'd served our apprenticeship, over the years tackling a total of six renovations, before moving on to larger developments - barn conversions. We now had two conversions under our belts, and were busy looking for our next project.

Thanks to modern farming methods, there are many redundant agricultural buildings disintegrating in farmers' fields, however, converting one of them into a home can be costly and fraught with difficulties. In most cases they can be in a fairly dilapidated state of repair: almost always needing major structural work to comply with current Building Regulations. The conditions imposed by Planning Departments throughout the country can also be onerous, and cause severe headaches for the poor developer. There always seems to be this tug of war, between trying to preserve the character of the original building, and creating a home fit for this century. But a barn conversion can be full of character and history, with all the comforts that modern living demands: a totally new house within the shell of an ancient building.

Gordon, a carpenter by trade, was well able to tackle the majority of the work himself but, with the larger renovations or barn conversions, we always used other skilled tradesmen - such as bricklayers, plumbers, electricians and plasterers. I chipped in with whatever needed doing; from ordering

4

materials to labouring although, as an interior designer, was better able to put in the finishing touches.

From necessity, we did our renovations and conversions the hard way. A relatively simple renovation would involve setting up camp in one room: trying our best to keep out the dust and rubble that would inevitably accumulate. There might be a sheet on the floor, a table for the microwave and, almost always, a bowl for the dirty dishes, that would need to be filled from a stop-tap outside. Fairly basic to say the least.

Barn conversions, however, were a whole new ballgame. These involved months, if not years, of hard labour; hence a certain degree of comfort was called for, in the form of a caravan or mobile home. Caravan dwelling on a permanent basis certainly leaves a lot to be desired, especially in the depths of the British winter *and* with a couple of dogs on board. While the conversions took place, our world revolved around the work: comfort being something we had to do without. As soon as the property was completed, if not before, we would move in; it would immediately be put up for sale and we would start looking for our next project. If we were very lucky, this would possibly give us six months, living in supreme comfort and luxury in the newly converted property.

Taking us completely by surprise, our last barn conversion, in South Devon, had sold on the first day it was put up for sale: amazingly at the full asking price. We had only just breathed a heartfelt sigh of relief at completing the work; not even begun to consider where we would go from there, let alone started the hard slog that involved finding just the right scheme to commit ourselves to. For us, there always had to be a compromise. We needed to find a property at the right price, so that we could make a respectable profit on the project on completion but, as it was our home, it also had to be somewhere we would enjoy living, however briefly we were left to enjoy it. So when we received such a good offer for our latest barn, we packed our bags, sadly said goodbye to Devon, and went to stay with relatives in Shrewsbury for a few weeks, before moving into a rented house nearby.

We'd been living in Shrewsbury for almost four months and were starting to get pretty desperate. Spring was staring us in the face, and a longing to return to Devon began to overwhelm us. Our rented accommodation was comfortable enough, but we were getting bored and fractious, with little to occupy our time. We'd been looking around the Shropshire area for something suitable; however, nothing we'd seen had grabbed our attention. Whether this was something to do with the fact that the weather was wet and dismal, or just that we hankered after the Devon countryside, I'm not sure. We just got grumpier and grumpier and our two dogs, Tess and Sharp, got muddier and muddier.

No doubt fate was to play its own small part but, just in case it needed a helping hand, we'd been having the Western Morning News (a Devon and Cornwall newspaper) sent to us through the post. We were hoping that we might see a property of interest advertised, in an area we already knew very well. As we made a living from our conversions, it was important to buy a property at the right price and in the right location, to attract as many potential buyers at the end of the conversion, as possible. Knowledge of the area we would be buying in was all-important.

As soon as the property supplement arrived every Saturday morning we would pour over it, feverishly scanning it for a potential project. Then, one Saturday, we spotted a promising looking advert for a redundant barn on the outskirts of Exeter in East Devon. It was difficult to see precisely what was for sale, as the grainy photograph was quite small but the description, short as it was, sounded interesting. It told us that the complex consisted of a redundant stone-and-cob barn, an almost new Atcost agricultural building together with three acres of pastureland. There was no price guide shown in the advert, so Gordon immediately got on the phone to the agent to see if he could get a few more details, as well as an indication of how much it was likely to sell for.

He looked more than a little puzzled when he put the phone down.

'Mmmh. Can't make out what the problem is, but there definitely seems something a bit strange about it,' he said, slowly. 'The agent was keeping something back I'm sure.' A deep silence followed while he cogitated and digested the information.

'Well? What did he say?' I snapped, getting more than a little rattled at the lack of information being passed on.

'Seems that it's a Receiver's sale and there's no planning permission going with it. In fact, the agent wasn't sure of the planning history but thought that no permission had ever been sought. Hmmm, why not I wonder?' he pondered thoughtfully.'You'd have thought they'd have at least tried for planning permission first before auctioning, to maximise the value.'

'Perhaps it's been turned down and they're not saying,' I guessed.

'Could be. I'll certainly have to do some investigating, if we're really interested. I think it might be better to go down to Devon to visit the Planning Department in person rather than phoning.'

'But are we really interested?' I asked.

'I definitely think it's worth a look. What do you think?'

'Difficult one. If we don't go and have a look at it, we'll always wonder what we've missed.' I knew him too well.

'Very true. Problem is that the auction's coming up the week after next. So ... when shall we go?'

After spending the whole morning discussing it, we eventually decided that he would have to view the property on his own. We had the complication of two large hyperactive dogs to consider. One of them was a working Collie called Sharp that we'd acquired along with our last barn conversion. The farmer was retiring from farming and had bought himself a posh new bungalow, where the dog most definitely did not fit into the scheme of things. Poor old Sharp had been left tied in a barn and not fed for a few days when we found him. At the age of ten, the farmer considered he should have done the decent thing, and long since popped his clogs. Sharp had other ideas! Although very emaciated when we first took him on, he was still lively enough now despite being twelve-years-old. Our other dog was called Tess, a beautiful tricolor Collie-Retriever cross we had brought up from a pup. Both of them were good dogs and reasonably well behaved but, together, quite a handful. I was disappointed not to be able to see the property but it seemed much easier for Gordon to go down to Devon on his own, and me to stay holding the fort.

He left two days later, with a huge list of questions we would need answering, before we could consider bidding at the auction. His first priority was to view the barns with the estate agent. He told me later that the estate agent was no more forthcoming at the viewing, than he had been on the telephone. What the agent did say, was that he was acting for the Receiver, and knew very little of the history of the barn. When buying property, we always make enquiries locally, therefore, Gordon took the opportunity to visit the immediate neighbours, to find out what he could from them. He told me later that he had met one of the farmers who lived in the tumbledown old farmhouse opposite and had asked him about the planning history of the barn. The old farmer, whose name turned out to be Ted, didn't seem to know anything other than that it was built at the same time as his farmhouse: sometime around the sixteen hundreds. He said that he'd found Ted's accent difficult to understand and his attitude a little bizarre given that he had lived next door to the barn all his life and yet knew nothing about it, not even who owned it. Or did he? Gordon gained the impression that he knew more than he was letting on.

While he was in Devon, Gordon took Dave Goodenough to see the barn. Dave was an architect and structural engineer we'd used on many of our projects. After viewing the barn he was very impressed with the scheme, declaring that he'd be more than pleased to draw up the necessary plans for us, just as soon as the barn was ours.

Time was also made to visit the Planning Department in the local town and he discovered, completely to his surprise, that no planning permission had ever been sought in the past. The Planning Officer was very positive and gave him the impression that, if suitable plans were submitted to the committee, he couldn't see why they shouldn't be passed: he could foresee very few problems. It looked as if we might have found our next project after all.

Gordon was full of enthusiasm when he returned from Devon, and had a head brimful of plans for converting the barn into a fabulous home. There were actually two barns being auctioned in one lot. One was an enormous modern steel-framed agricultural building that we could smarten up and use for storage, as well as garaging. The other was built of cob; a material we hadn't worked with before. Cob is a strange substance, which has been very popular for building all kinds of structures down the centuries. It's a mixture of clay, straw and small stones, moulded together to form a very tough building material. This barn was at least one hundred feet long, in fact probably too large for us, but would be a magnificent space in which to create a home.

Our rented house was littered with drawings, floor plans and odd jottings. Everything seemed set fair, just as long as we managed to outbid everyone else that is. The wait until the auction was an anxious one for both of us.

The day of the auction dawned at long last and Gordon set off in the morning for the venue in Exeter. It was timed for the early evening and was to be held in an hotel just outside the city centre. By the time he left Shrewsbury we still hadn't heard from our solicitor, who had been making various legal enquiries about the property. Apparently, he had been having problems getting answers to his queries but, by lunchtime, had telephoned me to say that all was in order.

The auction was delayed for more than an hour because the organisers had, at the last minute, decided to call in a security company. It seemed that they had been half expecting some kind of disruption from the bankrupt owners and wanted to make sure that they didn't attend the sale. They had also arranged for the whole proceedings to be filmed – just in case. As it happened, there'd been no trouble at all, although the auction house had been packed with locals, all expecting a punch-up and to see the fur fly. According to snippets of information that Gordon overheard, or been told at the auction, the previous owners of the two barns were an old local family, named Luscombe. There had been much talk of gambling debts and bank loans that had, eventually, ended with them having to sell most of their land and farm buildings. People were anxious to tell him tales of past feuds: he got the impression that most of the locals were wary, if not actually frightened, of the Luscombes. More than one man had told him that it had been more than his life had been worth to bid for any of the Luscombes' land or barns.

Gordon told me that because of all the expected trouble, there had only been one other man bidding against him. He was also a stranger to the area, and hadn't known of the problems involving the previous owners. In any case he only wanted it for a cattle store and, as soon as his low estimate has been reached, he pulled out of the bidding. As our intention was to convert it into a house, our budget was much more generous and, with no competition, he was totally amazed when the hammer went down on his last, very low, bid. It took him a few minutes to register exactly how low the bid had been and just how lucky we were. *Or were we?* Gordon also found out that the previous owners of the barns still lived opposite in the dilapidated old farmhouse and - like it or not - we would be neighbours. Neighbours from hell, some locals had told him after the auction and we should be extremely careful when dealing with them. We were committed now and had no choice other than to make the best of the situation.

The next month was spent packing our possessions and looking for a caravan fit for us to live in while the barn was being converted. Because the conversion would be a long one, we had at first considered a mobile home but, as the approach lane was very narrow, Gordon thought delivering a mobile home on a trailer would be too problematical. There was also an old bridge just before the barn to negotiate, not in the very best state of repair, and an unknown quantity as far as load bearing was concerned. Our imagination ran riot thinking of the nightmarish scenario of a monster delivery lorry, topped by a mobile home, either getting stuck half way across the bridge, or else the bridge collapsing totally under their combined weight. We took the easy option, finding a fabulous brand new, twenty-three feet long, twin-axle caravan at a bargain price, which was comfortable enough to be our home for the next few months. Our furniture was taken away into temporary storage, just leaving us with our clothes and one or two treasured nick-knacks. Gordon was planning to eventually construct some sort of storeroom for the furniture, in the modern Atcost barn, just as soon as time permitted but, for now, we would have to bite the bullet and pay a weekly storage fee to the removal company.

A few weeks later, at the crack of dawn, we packed the caravan to the gills with our bits and pieces and set off for Devon in high old spirits. The long trawl down the motorway was interminable and I spent much of it either asleep or dosing, leaving long suffering Gordon to do all the driving. After leaving the monotony of the motorway, I began slowly to realise that we were passing through some lovely countryside. The trees were just starting to bud with their first pale green leaves and the grass verges were packed with golden daffodils. While I'd been sleeping, the concrete and brick had turned into green fields and leafy lanes. I had forgotten, during my sojourn in Shrewsbury, just how lovely Devon could be. All tiredness forgotten, my excitement began to mount, as we drew nearer to our destination.

Suddenly Gordon said, 'This is it, just along here on the right.'

I leaned anxiously forward waiting to catch my first glimpse. We pulled off the road and came to a stop at the top of an overgrown narrow lane, barred by a metal gate. The lane led down into a dip with a cluster of very dilapidated buildings huddling at the bottom. Climbing stiffly out of the cab to open the gate, I gazed around me in bewilderment: I found it very difficult to take it all in. On the right hand side of the lane was a decrepit old farmhouse, with a large and distorted multi-paned window, jutting out precariously just under the overhanging, very patched, grey slate roof. In front of the farmhouse was an equally decrepit shippon, or cattle shed. On the left hand side of the lane were the two barns that now belonged to us. One of them was an enormous two-storey cob barn that was almost as large as the farmhouse. Fronting the lane, it was long and narrow with an incredible wavy roofline. It disappeared into the distance, behind a large tree that stood in the front garden of the farmhouse. Behind it, and at right angles, stood a huge modern agricultural building with the maker's "Atcost" sign on the front. An Atcost building is one of those ubiquitous pre-cast concrete and steel structured buildings that seem to sprout from every farmyard throughout the length and breadth of Britain. Common as mushrooms on a cow pat, but very useful all the same.

All the buildings were grey and dismal looking with the only splash of colour coming from the very green lawn and massive tree in front of the farmhouse. My high spirits of a few moments before took a distinct nose-dive. A cloud as grey, and as heavy, as those just forming above in the sky, settled over my head, as I looked upon this truly depressing scene. This was Cold Comfort Farm indeed. It all looked as if it had been left in a time warp. Any minute now sixteenth century farm workers would be bringing in the cows for milking. *What had we done?*

Whistling softly and looking as if he hadn't a care in the world, Gordon drove on down the lane. On the right we passed the shippon with its sad sagging roof. The cracks in the old lime render, on the front- and side-walls, were wide enough to put your fist through. The shippon had a fenced-off yard in front of it, slick with muck, and there was a neat heap of old dung in the middle that must have taken quite a few months to gather. The yard and cow shed looked totally deserted, as did the farmhouse beyond. Perhaps the ghostly cattle that had been housed there had left the muck heap in a previous century?

We turned into our entrance on the left, pulling up before another metal gate. The barns were surrounded by a picket fence with a dung- and straw-strewn yard between the two. The gate we were parked in front of led into this inner yard. At the far side of the inner yard, attached to the Atcost, there was yet another metal gate with the fields and orchard beyond. I spotted a tiny caravan just inside the field, leaning drunkenly against the other end of the Atcost building. We could have saved ourselves the trouble of towing a caravan all the way from Shropshire, if we'd known we were going to inherit one!

I clambered out of the Land Rover, swinging wide the gate, as Gordon drove through into the enclosed yard. Glancing casually around, I did a double-take. To my utter horror, at the far end, under a partly demolished lean-to, an enormous bull was standing watching me. He was big - very big - and had been feeding from a straw bale, standing on the floor by his side, until being rudely interrupted. He tossed his broad shaggy head in the air, snorting disdainfully, assessing me with his small beady eyes. It didn't take me too long to decide that I needed to be out of there - and fast. Running is normally a pastime I avoid like the plague, but this situation called for desperate measures! I ran as quickly as my quaking legs would carry me to the final metal gate, attached to the side of the Atcost. This was firmly bolted and, in my haste, I couldn't seem to get the rust-encrusted bolt to draw. I debated whether to just climb over it, abandoning Gordon to his own fate, but thought better of it. Not daring to look behind me, I fumbled and bumbled, sawing the bolt to and fro in its socket, until it at last gave way in my sweaty hands. Dragging the gate open, I waved the Land Rover and caravan through, clanging the gate shut behind me, with an intense sigh of relief. *Phew, that was a close call.*

But my relief was not to last long: farther on up the yard, I could see three goats galloping towards me like maniacs. *Oh hell, what have I done to deserve this?* I hastily clambered back into the Land Rover, slamming the door with more force than the driver approved of. The goats milled around us for a while, giving me plenty of time to observe that one of them had a very wicked set of horns which I had no intention of getting too close to. Gordon was, by this time, roaring with laughter, assuring me that they were harmless. *Yeah, right.* How on earth was I going to live here with a herd of goats roaming the place sporting horns six miles long? Not forgetting, of course, our friend the mad bull to negotiate every time I wanted to pass through the yard. This little escapade was turning into an absolute nightmare. I didn't like it here one little bit.

Fortunately, the goats soon lost interest and wandered off. I set off very timidly to explore; keeping a wary eye out for the goats and anything else that this awful place might want to throw at me next. I thought of myself as an animal lover, but hadn't had too much to do with anything larger than a dog and, right at that moment, didn't want to - thank you very much.

We had parked the Land Rover and caravan outside the Atcost building. Gordon had already earmarked this incredibly large building as the place to park the caravan inside. Here we would have a degree of protection from the still bleak March weather. He also wanted to build some sort of wooden enclosure, inside the building, so that there would eventually be somewhere clean and dry to store our furniture. There was also some discussion about setting up an office desk and telephone. It would, all in good time, become a regular home from home; our campaign headquarters no less.

Making my way across the slimy dung-encrusted yard, towards the huge metal doors, I could see that it was divided into three sections, by block walls, topped with metal rails. Each section would be amply wide enough to cope with the caravan - if it had been empty that is. In the far left-hand section there must have been about thirty bullocks, snorting their foggy breath into the air. They give the impression they were standing quite high off the ground but, on closer inspection, I could see that the straw and dung they were standing on, had gradually compacted over the years, and now left them a couple of feet in the air. The middle section was relatively clear, only having a few rusty old farm implements and a colossal pile of rotten timber cluttering it up. The right hand section was stacked to the roof with bales of hay and, tucked in amongst them, was a newly born lamb and her anxious mum. The barn was supposed to be empty!

When Gordon had returned to Shrewsbury from the auction, he'd told me that when he'd viewed the barns with the agent, there had been a great assortment of animals in the Atcost, in the cob barn, and grazing on the land. Just in case they should be forgotten about, he'd asked our solicitor to request that the Receiver remove everything that was *not* included in the sale before completion. Obviously it had slipped someone's mind, or we'd bought more than we bargained for!

We were both very tired after our long day and could have done without these added problems. It was, by now, almost dark: there was little to be gained by doing anything other than getting something to eat and then trying to sleep. Hopefully, things would look a little less bleak in the morning.

We both had a good night's sleep and, after making a phone call next day, we were in a more positive mood. Our solicitor had promised that the Receiver would be contacted and he would ensure that all the animals would be removed that very day. Not only that, but someone would also come along and remove the hay so that we could put the caravan inside.

While we were waiting, there was plenty for us to be getting on with. The yard, at the back of the Atcost building was a mess, with animal droppings spread liberally all over the place. As we had two dogs that were getting very smelly trotting through all this muck, we thought it the better part of valour to make this our first job. Just in case we should need anything else to occupy our day, there were also hundreds of old tyres and bits of black polythene littering the land, which would also need moving. We set to cleaning up with a vengeance.

Our first priority was to clean up the muck. This was very generously spread over the whole surface of the yard (an area of about fifteen-hundred square feet), plus an enormous pile of it that must have been at least six feet high. Gordon decided firstly to dispose of the pile, by shovelling it into the trailer and then towing it up to the top of the field, where it would eventually break down and rot into the ground. After shovelling for a few hours, we suddenly exposed a hoof. Further investigations, with a gingerly wielded spade, revealed that this pile also contained two rotting sheep carcasses. *What* a nice surprise! Quickly covering the stiff bodies with some of the black polythene that was lying around, we decided to leave them to lie where they were for the time being.

It had been a hard day. After a quick shower each, we sat enjoying a rare March sunset, complete with a glass of red wine apiece. The beauty and potential of our new home were just beginning to dawn on us: this was indeed a lovely spot. The land around us rose upwards from the yard into an orchard and, although the trees had obviously seen better days, they would fairly soon be thick with creamy pink and white blossom. We had our very own apple orchard. Discounting a few of the obviously rotten trees, there looked to be in excess of fifty that would probably yield a good crop. There was also a stream, meandering down to a small bridge, which partly dissected the land. The banks were quite high on either side of the stream, liberally sprinkled with pale yellow primroses and clumps of tiny violets.

We sat enjoying our wine, while a flock of hens and vividly coloured cockerels strutted around our feet, trying to find the odd crumbs we had left from our scrappy dinner. We could see the goats on the other side of the stream, chomping away at any old branch or bit of tough looking root that came their way. Thankfully they had completely ignored us after their first rush of curiosity. The air was scented with wood smoke: someone was burning aromatic apple wood. We might have a long way to go but we could both see that one day this would make a fabulous home.

CHAPTER TWO
INTRODUCTIONS

We slept the sleep of the truly exhausted; waking at the crack of dawn, refreshed and raring to go. I let the dogs out of the caravan, after lazily getting dressed in yesterday's filthy clothes. I stood in the doorway for sometime, drinking my tea and admiring the view. It was cold: there had been a bit of a frost overnight, which was rapidly dissolving in the thin sunshine, leaving clumps of low hanging mist over the orchard. The goats were already foraging for breakfast; grazing precariously on the steep slopes of the stream bank.

What was that? I caught sight of a shadow at the far edge of the new barn and set off bravely to investigate. There were so many hazards in this place that I began to feel like an intrepid explorer. My intruder turned out to be a large thickset man of indeterminable age, wearing a thick navy greatcoat, which looked as if it had first seen the light of day in the trenches of Normandy, rather than the green fields of Devon. It was button-less, tied with orange binder twine, wound around his comfortably ample middle, and tied with a flourishing bow at the front. His hair was extremely wispy and sparse, barely covering his brown freckled pate. His Wellington boots had obviously seen sterling service for many years, and were caked with much more than just mud. Our dogs, who were excitedly sniffing around him, seemed, not surprisingly, particularly interested in those wellies. As

I drew closer to him, I caught the first blast of the very pungent smell that emanated from him, a signature smell that I would get to know all too well in the months ahead. It was extremely unpleasant. Trying not to wrinkle my nose in disgust, I said, 'Good morning. What can I do for you?' As far as I could make out he said 'mumble, mumble, mumble.' *What on earth had he said to me?* 'Sorry what was that?' I asked.

'G' mornin' to ye,' he mumbled again, and this time I could just about decipher what he said. He grinned at me showing his one tooth. 'Mumble, mutter, mumble,' he continued.

'Pardon?' He didn't seem to have any more than that one central top tooth, and the words he spoke rolled too loosely around his gums for comfort. After a great deal of intense concentration, I could just about make out the gist of his opening gambit.

'Jus' seein' t' bullocks, like. 'Ansome mornin' 'int it?'

Noticing the wisps of hay still adhering to the front of his coat, I came to the conclusion that he had just been feeding the cattle. He was now dipping his ham-like hands into a rusty paint tin, preparing to scatter corn over the yard. This was, presumably, for the poultry that were squawking around his ankles.

Penny dropping, I replied, 'Ahhh, you must be from the farm opposite. I'm Maggie, your new neighbour.'

'Eh?' He mumbled, cupping his hand behind his ear. He was either hard of hearing, or else my accent was posing the same problem for him, as his Devonian burr was for me.

'I'm Maggie,' I repeated.

''Ow do maid, I'm Ted.' His round, ruddy complexioned face, split from ear to ear in a gummy grin.

We stared at each other in silence for a while. No doubt I looked as strange to him as he did to me! He obviously expected more of a conversation than he was actually getting so, searching desperately around for something to say, I began with a subject that had been bothering me in the night. 'Do you know who owns that little caravan on the other side of the barn?'

'Dun rightly know, I'm sure. You must ask our George, he'll be the one to know.'

'Right, I will; only problem is...I don't know who George is!' I replied.

'Why he's me brother o' course! There's George; he's the eldest. Then there's me. Then there's our James and then comes sister. She's our Verity. Our George is the one to ask about anything you wants to know.'

'OK, I certainly will. Have you just been feeding the cattle?'

'That *is* right. Does it every mornin'. I'd best be gettin' on then. We'em off to market soon,' was his parting shot, as he ambled his way back across our yard.

When I got back to the caravan, Gordon had already had his breakfast, and was in the process of putting on his working boots.

'Hey, guess what? I've just met a strange old chap called Ted. I found him wandering around in the yard.'

'Oh yes? He was the one I spoke to before the auction. He's a good old boy but his accent is a bit hard to understand. How did you get on?'

'OK, I think, although we didn't really know what to say to one other. You're right; he is hard to understand, I felt as if I was constantly saying, "pardon". He said he was feeding the animals but, if they're due to go today, then he won't be doing that for much longer. It was a bit weird seeing him just wandering around our barn. And the smell was enough to wipe out an army at a thousand paces!'

Gordon started to chuckle, 'Yes, he does whiff a bit doesn't he? The smell takes a bit of getting used to but he's OK really. Ted seems to be the spokesman for the whole family. Whenever I came to see the barns he was always the one hanging over the gate and, if I did manage to catch sight of any of the others, they would always seem to disappear a bit sharpish. It strikes me that they send him out to see what's going on,' pondered Gordon.

'Well, he's pleasant enough but I'll be glad when he's not checking up on us. Can we go and see the cob barn?' I asked. At that point in time, I hadn't even walked round the outside of it yet, let alone seen inside.

'Oh, course! Let's go before we start clearing up.'

And here was I forgetting about the bull! As we got to the gate I could, thank heaven, see the lumbering mound of him still tucking into the hay bales at the far end of the yard, under his crumbling lean-to shelter. I stood, dithering, at the gate.

'He's no problem, I don't know what you're so scared of,' Gordon said, correctly interpreting the look on my face.

18

'He's built like a brick shit house, why else do you think!'

I will admit to surreptitiously watching what the bull had been up to from time to time. I mean, how quickly do bulls move anyway? The ones I'd seen, when I'd caught snatches of bullfights on TV, seemed to go like the clappers. This Devon specimen seemed more clapped out. He only appeared to saunter from one place to the next, as if his feet hurt him and he needed to change into his slippers. I had also seen him with his heavy head on the top rail of the fence, just gazing lazily around. He looked shabby and old. Perhaps I was making a mountain out of a molehill. But bulls are bulls and, I'm the first to admit, I know nothing at all about them.

Gordon had already opened the gate and had started to saunter nonchalantly across the yard, without the bull moving a muscle. Taking another swift glance at it, I took to my heels and dashed across to the other side, rapidly letting myself through the far gate. I was going to get very fit at this rate. There was no way I would ever get used to *that* animal being in my yard. What if he was nearer the gate next time or, horror of horrors, if he was actually at the gate itself? The only other way to get back to the caravan was through the field that ran along the far side of the stream, and then through the orchard, usually containing the beasts with many horns (the goats!). What an alternative!

Putting aside my anxieties for the present and following Gordon up the lane, I could see that the cob barn was more than a hundred feet long by about twenty feet wide. The front of it perched along the side of the lane, much lower at one end than the other. Standing back into the lane to get a better overall view, I could see that the end nearest the yard gates had a small window near the apex of the roof, which must mean that a section of the barn had two storeys. It was difficult to see where this second floor ended, as it appeared to be the only proper window opening on that upper floor except for a couple of arrow slits on the front facade facing the lane. The far end of the barn didn't look tall enough to have two storeys, as the whole building sloped dramatically. There was a great deal of earth banked up round that far end, which could well mean that the floor level there was under the level of the road.

There was a small porch jutting out from the barn, about half way along its length. The door was not locked and when we got inside, we could see that this little space used to be the old dairy. There was still a substantial stainless-steel tank right in the middle of the floor, probably used for milk storage, with various rubber tubes sticking out of the side. The walls were covered in telephone numbers and jottings about the milk yield of each cow. Peering at the list in the gloom, I could just about decipher that the herd had only consisted of eight cows, with quaint names like Buttercup and Tinkle.

We came out of the porch and strolled further up the lane. We passed quite a few large openings at ground level, mostly boarded up with old timber. Good. These were absolutely essential for letting light into the barn. I knew, from past experience, that the boffins in the Planning Department were never keen to let you create new openings, so the more that existed, the better. When there were insufficient openings to create enough light for modern day needs, the Planners seemed, for some inexplicable reason, to prefer the use of Velux roof lights. Even so, I could see that we would probably end up with quite a few of those roof lights, as there were very few openings at first floor level. All the openings we had so far found had been either padlocked, or nailed shut, so what was on the other side would have to remain a mystery until later.

At the very far end of the barn, there was a set of moss covered stone steps, leading up to a door set just below the roof. We started up the steps but halted half way, when the sound from inside finally alerted us to the fact that the barn was evidently not empty. The old rotten timbers of the door heaved and shuddered with the onslaught of bodies hurling themselves against its planks. It sounded as if there was a huge pack of dogs inside: we could hear them snarling and snapping. Even had we been foolish enough to want to open it, the door was in any case barred with a padlock and chain. I was quite clearly going to have to wait a while before I saw inside the building. *Oh well, explorations over and back to work.*

We worked hard again that day. There was one thing for sure - we were not going to be short of things to do for quite some time. Just for starters, behind the Atcost barn, there were literally hundred of tyres. Tractor tyres, car tyres, trailer tyres, you name it they were there. Gordon told me that they were probably used for piling on top of the silage, to hold the black plastic sheeting down. He was probably right. There was enough black plastic lying around the place to warrant that theory.

It was great fun at first. Gordon stood at the top of the grassy bank, where the largest pile of tyres lay. He started to roll them on down the bank towards me. I was standing on the large concrete area at the back of the Atcost barn. I would then give them an extra roll to send them onto a pile, just off the concrete, but before the land dipped down to the stream. Just like rolling hoops with a stick! Timed right, they would end up in quite a neat heap. Timed wrongly, they went into the stream.

It got a bit monotonous after an hour and my arm had begun to ache. After two hours, things were getting serious. My back, arms and neck were now giving me serious grief. I could see that I needed hardening up - and fast. It was definitely a good idea to pack up for a drink and a bit of a breather.

At the end of the day we had just about moved all the tyres. There were so many of them that it was impossible to count, let alone estimate, how many there were. I knew the tyres were going to be very difficult to get rid of because I'd already phoned the Council to ask where we could dispose of them, and they'd told me they wouldn't take them locally. We eventually found out, after many phone calls that the only legal disposal depot for tyres was in Bristol, more than eighty miles away. We had discovered yet another job that was going to have to wait for another day.

Frustrated in our efforts to complete at least one simple task, we decided to tackle the other job we were trying to avoid. The two rotting sheep carcasses from the dung heap were given a decent burial at the top of the orchard, and the rest of the muck from the pile was quickly scattered on the field around them. The yard behind the Atcost building was then hosed and scrubbed to within an inch of its life. It was rewarding to finally complete one mission. We would now be able to walk around this area without leaving a trail of evil smelling sludge behind us.

We had set up the camping table and two chairs outside the caravan and here we sat enjoying yet another fabulous sunset. It had begun to get cold by this time, but we were still revelling in the great outdoors; determined to stick it out for at least another half an hour before we moved inside to get dinner ready. We chatted aimlessly away about the day's events. I could see the goats still over the other side of the stream - best place for them. I felt extremely relieved they'd stayed in their own territory. Then I spied the cocks/hens crowd milling away at the other end of the concrete apron.

'I thought all that lot were supposed to have been removed today?' said I, gesturing towards them.

'Oh, hell! I'd forgotten totally about the animals,' Gordon replied.

'Well, either someone has forgotten to collect them, or else no one wants them. Oh...and what are we going to do about the other bits and pieces: that little caravan and the trailer? And I was forgetting the cattle crush?' I reminded him. A diminutive caravan had been left leaning drunkenly against the Atcost barn wall and a small trailer, with a blue tarpaulin cover, tucked away round the back of it. There was also a cattle crush (used when large animals need injections or to be inspected) left in the yard between the two barns. We had no more idea who owned these articles than the animals we'd been left with.

'I suppose that means going down to the phone box again tomorrow to chivvy them up. What a waste of time,' Gordon shook his head. 'I can see I'm going to have to lay the law down and tell them if they're not moved, and pretty damn pronto, then I'm going to sell 'em.'

The visits to the telephone box, about half-a-mile down the road, continued almost daily with Gordon getting more and more irate. It wasn't until three weeks later, that we observed an antiquated cattle truck arrive at the top of the lane. It eventually made its laborious way down the track and stopped, with a juddering cough, in front of our yard gates. It looked as if here, at long last, was someone to take away a few of our problems. It turned out that the men in the truck had come to collect only the bullocks. They had been renting the barn space for the winter period from the people at the farmhouse and, until our solicitor had contacted them, hadn't known that the barn had changed hands. They were still paying rent for the bullocks' storage and keep, and thought they could leave them in the barn for another couple of months at least, until they were ready for market. They were not best pleased at having to find new quarters for their beasts, and a bit of an argument started between one of them and one of the Luscombes. All must have eventually been sorted to their satisfaction as hands were shaken, the tailgate of the lorry was quickly lowered, and their eagerly pushing and shoving beasts loaded into the truck. They slowly pulled away up the lane, belching exhaust fumes, and out of our lives.

That left only the ewe and its still tiny lamb, the goats, the bull, the cocks, the hens and, last but not least, the rowdy pack of dogs in the barn. One down, six more to go then. *Fine...*

More phone calls were to take place during the next few days, but no information emerged concerning who owned the rest of the motley crew. I guessed that Ted was still wandering around on our land during the early mornings, despite the fact that Gordon had asked him not to. The bullocks had now departed for pastures new but his excuse still seemed to be that he was feeding the rest of the animals. Well, perhaps he might know where we might find their owner. I decided to tackle him and find out.

Spotting him early one morning coming out of the Atcost barn, I hailed him. He turned round and waited for me to catch up.

'Hi Ted. Nice morning!' I said as I approached him, hurriedly taking a step backwards: it wasn't wise to get too close. The smell was appalling today, a bit like those sheep we'd had to bury. 'Do you know who owns all these animals that have been left here? Someone came a few days ago to collect the bullocks but we seem to be left with a lot of others that no one seems to want to claim.'

'Eh? What's that? What's that ye say?' He said irritably.

I repeated my question more slowly. He mumbled back at me and I said my usual "pardon". He took a long look at me and repeated slowly, and this time more clearly, 'Why they'm ars o'course. That's why I be a feedin' em, maid!'

'Well you just can't leave them here, Ted,' I reasoned. 'The barns have been sold to us now: you really need to think about moving them onto your own land, don't you? We want to get our caravan in the Atcost barn and we'll be pulling the cob barn to bits very soon. Can we move at least some of them out today? Please Ted? Especially that huge bull. He frightens me to death every time I have to cross the yard.'

'I dunno about that maid, where'll us put 'e?'

'You'd know the answer to that one better than I would surely, Ted. What about the shippon in front of your house, can't some of them go in there?' I pleaded.

'Nooo...o'course not. Thur be a couple of beasts in the field over yonder that go in that thur shippon, when the ground gets wet, like. Us 'ull 'ave a little think on, and see what us can do,' was the only answer it seemed I was going to get that day. He shuffled off, leaving behind him the piquant notes of his extraordinary aroma.

Neither Gordon nor I saw anything of him for a couple of days. He was obviously lying low. Still "thinking on" no doubt. Gordon eventually spotted him in the lane and, not wanting to let the matter drop, set off to tackle him again. He returned having got just as far as I had – absolutely nowhere. Gordon had been given a verbal run-around, with Ted saying they were going to make arrangements soon, for somewhere else to move the animals to. Ted did, however, give us a clue about the caravan and trailer. Apparently they belonged to a friend of theirs who would be moving them "very soon". *Likely tale!*

That evening we sat relaxing in the camp chairs with our usual glass of wine trying to resolve the problem of what to do with the remaining animals. We seemed to go round and round in circles, within an ace of finding the right solution, but never quite getting there. We didn't want to step on too many toes, or upset our new neighbours, unless there was absolutely no alternative, but at the same time, didn't see why we couldn't just be left to get on the with job and enjoy our new home in peace. The only conclusion we finally came to, was to let things lie quiet for a few days, and see what action Ted or his brothers would take.

We were obviously misplaced in thinking action would be taken - because it wasn't. Two weeks went by and not a birdie. Not a sight nor sound of any of them. Were they hiding from us? Eventually, enough was enough. Gordon, in high old dudgeon, marched up to the back door of the farm and knocked loudly.

There were two entrances to the farmhouse. One led from the lane, through a small metal side gate, into the garden and thus to the front door. This gate was, as far as we could ascertain, always kept fastened with a substantial padlock and chain. We had also never yet seen that massive metal studded front door open. The other entrance was at the back of the farmhouse, through a walled yard. This was strewn liberally with unidentifiable bits of metal, rubber and drums of this and that. There was a carport facing this yard, built of unrendered blocks laid higgledy-piggledy, with a corrugated-iron roof perched on top. Inside there was a rotting sofa with its horsehair liberally exposed, together with other bits of broken furniture, supporting a family of a dozen or so cats.

Just inside the entrance to the carport, it was usual to see at least one dog tied up by a piece of fraying rope. The rest of the pack of dogs - perhaps six or seven - usually roamed around the yard or land that adjoined it. It was pretty difficult to see exactly how many dogs there were, because they were all more or less identical. They were spindly looking things, with dull grey-brown coloured coats and sharp mean looking faces. The breed was definitely a mixed one, but contained at least a little whippet somewhere along the line.

Usually the one tied up was a bitch in season. The actual reason for it being tied up was a mystery to me. I was under the impression that unless you wanted a litter of puppies, then a bitch in season should always be kept away from the male of the species. Presumably more puppies were wanted as, from what I could see, every dog in the pack was definitely doing its level best to provide the inevitable pups. Not only that, but from the level of noise, there must have been another six or seven dogs, kept in the top end of the cob barn, that we hadn't yet seen. What on earth did they do with all those dogs?

The dogs set up a terrific row when Gordon entered the yard and I could only admire his daring in standing his ground, waiting for the door to open. They milled around his legs in a hostile manner, growling and showing their yellow teeth. Cowardly me had stayed the other side of the gate. After a long wait, while Gordon's temper simmered gently, Ted eventually opened the door, "Ow do boyee,' he said.

'Ted, can you come and help me shift some of these animals,' Gordon barked, 'I want to clear at least the Atcost barn today. I need to get that ewe and lamb out now...so where can we put them?'

'Dun rightly know, I'm sure,' Ted said, scratching his almost bald pate and stalling for time.

'Well...they have to go somewhere today, and that's for sure! You'd best think a bit sharpish, because my patience is fast running out,' Gordon said forcefully.

Ted's face lit up as if a sudden revelation had come to

him "Ow 'ud it be if we put 'em in that there barn,' he said, pointing to our cob barn in front of him. 'That 'ud be a proper job an' no mistake. They'd be nice un dry in there. Us can't turn out old mother and 'er little un in the wet now, can us?'

Oh hell, he was going for the sympathy vote now!

'No...no I suppose not. OK, then we'll do that, for now. Come and give me a hand will you?' Gordon, cornered, had no other option open to him.

The anxious ewe and her bleating lamb were very soon transferred across the yard, and safely installed in their new quarters. Ted had taken the massive lock off one of the boarded entrances and I was able to see that this was once the milking parlour. There were still iron partitions and cattle feeding troughs where the cows would have been milked. All this was liberally covered in hay, very old dung and draped with lacy cobwebs. It was very dark and dusty inside but the little family would be cosy enough, with the clean straw we had put down.

'That's a job well done. Now how about moving a few of the others,' said Gordon, pushing his luck.

'Oh ah, us ull come and 'elp move them there goats temorrer, sharp,' replied Ted, not giving an inch, and retreating back into his yard.

'Hey! No you don't!' Gordon had other ideas. 'That bull is going today. I'm starting clearing that lean-to he's in tomorrow and he is definitely going - NOW.'

Ted, recognising a superior force when he met one, came willingly enough to lead the ponderous bull, by the ring in his nose, into the yard surrounding the shippon. I held my breath, but he went very quietly, thank the Lord.

At least we had the Atcost barn to ourselves now. Well - apart from the hens and cocks that is. Oh, and of course, I was also forgetting about the bales of hay. There must have been about fifty or more bales, stacked high to the ceiling. These bales of hay gave the place a warm cosy feel; providing a much-needed barrier, protecting us from the wind whistling round the cavernous barn. I wouldn't be too worried if they stayed for a while. What we would need to get rid of though, and pretty quickly too, was the dung from the bullocks. It was starting to whiff a bit and that section of the barn always seemed full of flies. It had been there so long that it must have been a couple of feet deep in places, and very well trampled in. It was not going to be effortless to shift.

Taking the easy option, we decided to put the caravan in the section where the hay now stood. This was relatively clean at the moment and only needed the bales re-stacking at the far end, to give us plenty of room. By the end of the day, this was all done and, after a quick sweep up to remove the odd handfuls of hay that had escaped the bales, the caravan was moved inside. Although this made the inside of the caravan dark, it did make it a whole lot warmer, and would go some way to protect us from the wind and rain. We put a small brightly coloured rug in front of the door and set the table and chairs out alongside it. Very cosy. Our new quarters for the next few months were at long last finally organised.

Later on, while we were out in the lane clearing up some of the rubbish that surrounded the front of the cob barn, a car stopped on its way up the lane. We knew there were two bungalows further on up the lane and had seen the odd car passing by from time to time. The woman, driving the nice little red Peugeot, leaned out of the window to introduce herself. 'Hello, I'm Kathleen from the top bungalow.'

'Oh, nice to meet you! I'm Maggie and that's my husband Gordon,' I replied.

'We'd heard you'd moved in but I hadn't managed to catch up with you until now. How are you settling in?'

'Fine, thanks,' I replied.

'We're having a bit of a get-together on Saturday evening, if you fancy joining us. Nothing special, just a few friends and neighbours for a bite to eat and a few drinks,' Kathleen said.

'That's very good of you to invite us.' I glanced at Gordon, could see his nod of approval, so without further ado said, 'We'd love to come. What shall I bring?'

'Whatever you like...something savoury would be a good idea and maybe a bottle?'

'Of course. What time do you want us?'

'I've told the others about seven, is that OK with you?'

'Yes, absolutely fine. See you on Saturday then.' We both waved, as her car continued up the lane.

I hadn't a clue what to bake for Saturday and would have to give it some thought. Gordon suggested a quiche and, although I wasn't at first too keen on the idea of making pastry in the small caravan galley, soon came round to the idea. I was given the afternoon off on the Saturday to do the baking, and also to get ready for our first evening out since we'd been living at the barn. I couldn't remember where I'd packed my glad rags and had to do a bit of ducking and diving in some of the lockers under the bunks, trying to find something suitable to wear.

The quiche turned out much better than I expected and clutching this delicate parcel, along with a bottle of wine and some cans of beer, we wandered up the lane to meet our other neighbours.

There was a large noisy crowd already gathered in the kitchen, which fell silent as we walked into the room. Kathleen approached with her hand outstretched. 'Glad you could come. Let me relieve you of all that and Alex will get you a drink, before I introduce you all round. Right.... what will it be?'

'Beer for me please, Kathleen,' piped up Gordon. She raised her eyebrows at me in a silent question.

'And a red wine for me, please,' I said.

She was even shorter than I, probably less than five feet tall. A lovely friendly roly-poly of a woman. 'OK everyone. Let's 'ave a bit of 'ush! This is Gordon and Maggie who've bought the Luscombes' barn, brave people that they are! I'll introduce your nearest neighbours first...this is Alex, my other half.' Alex was almost as round as Kathleen, probably in his mid-fifties, with pepper-and-salt hair, dressed casually in jeans and sweater. 'Next we have Brian and Brenda. They live next door.' So...Brian and Brenda were the people who live in the first bungalow beyond our barn. They were a couple in their thirties: both looked friendly and proved easy to talk to. 'Then we have Rick and Sue. They live in the farm that immediately borders our land to the north.' Rick looked a typical farmer, thickset and ruddy faced. He was probably in his early sixties and his wife was a robust fair-haired woman of about the same age. 'The rest of this rabble are all my relatives...Rob, Dave, Sheila, Paul and little Linda over there.'

They all proved to be a lively bunch. The main topic of conversation inevitably centered around the only other immediate neighbour who had not been invited - the mysterious Luscombe family living opposite the barn. We were to learn many startling snippets of information that evening. Not one of the people present had a good word to say about the Luscombes. In fact, it was alarming to witness the degree of animosity that they all felt towards them. Only having met amiable old Ted so far, we had been lulled into a sense of false security, and had forgotten the many comments made at the auction about the brothers' viciousness and long-memories for a feud.

Brian surprised us with his intensity, saying that the Luscombes had made his and Brenda's lives a complete misery, making them wish, on many an occasion, that they'd never bought their bungalow. He didn't go into specific instances, but said that George and James had been abusive to them both, warning us that we should keep well away from them, especially when they'd been drinking. Brian said the Luscombe brothers were offensive enough at the best of times but, after a few drinks, they were almost uncontrollable, shouting and swearing and waving sticks around. Or worse – guns. *Guns?* Yes, guns!

Rick was more forthcoming. He had lived in the farmhouse bordering the Luscombes' land all his life. He was younger than them, and when he was a boy of school age, they were all young men. Whenever the opportunity arose, the Luscombes would bully him, even throwing cow muck and stones at him if they could. He said he recalled often waking up in the middle of the night, looking out of his window to see a tractor and flatbed trailer careering round the farmyard. The Luscombe boys would be hanging onto the back, screaming and shouting insults, all- guns-blazing, like a scene from a second rate B-movie. Despite this, he still had a certain amount of sympathy for them, in their present situation. He told us that the Luscombes' farm had been heavily mortgaged to pay off the gambling debts that James and George had accumulated. Apparently, they liked to bet on just about anything that moved, including horses and dogs, and weren't averse to the odd game of "spoof". The problem was that their income didn't keep pace with their considerable outgoings, especially as they still continued to gamble. The debts had been allowed to mount for some time, but the bank had finally had enough, eventually calling in their loan.

The noise level in the room was rising above the music on the stereo, while everyone tried to tell us just what had happened when the bank had finally attempted to recover their money. Apparently they'd received no response to their letters and so had been forced to take them to court. Both Alex and Rick had attended the court hearing, so knew at first hand that the bank were given permission by the judge to sell the Luscombes' farm. The bank promptly instructed an estate agent, who attempted to draw up sales particulars. The Luscombes had at first co-operated with the agent and had allowed him access to measure up the rooms on the ground floor. Problems started apparently, for some unknown reason, when he attempted to do the same with the upper floor. Story has it, that he was thrown down the stairs. Presumably he was not badly hurt but, in any case, must have been able to see enough, before his fall, to make up reasonable property particulars as Kathleen produced a copy for us to see. Upstairs there were, apparently, five bedrooms but no bathroom or toilet. Nor did any kind of sanitary facility appear in the details of the ground floor. *So, where did they do it?* Speculating on this knotty question left us breathless with laughter. It was quite common for old farming properties to retain their outside privy or "thunder box", sometimes housing two or three holes for use by the whole family at once, even if they did have inside facilities. But at the Luscombes' no one could even recall seeing a structure that could possibly house an outside toilet.

Not long after the estate agents unhappy accident, the Luscombes appealed against the court's eviction decision and won their case. It was based on the fact that the brothers held the debt alone. Verity, their crippled sister, who had no share in the debt, owned part share of the farmhouse, therefore, the farm could not be used to pay off the debt. Rick and Alex were again in court to see Verity on the stand. The Luscombes' case was facilitated by the fact that Verity was crippled and her performance in court gave the impression that she was also mentally retarded. The judge in this case was obviously sympathetic, ruling that they could all stay in the farmhouse while Verity was still alive. All their land, with the exception of the four acres, which surrounded the farmhouse, would be immediately sold. This included the two barns that we subsequently bought. As soon as Verity either died or needed full time residential care, then the farmhouse would also be forfeit and the bank could take possession.

Then Alex's voice asserted itself over the hubbub. 'Have you had the flashing light treatment yet, when you come in after dark?'

'Flashing light treatment? What's that in aid of?' queried Gordon.

'If you come home after dark, the Luscombes flash their outside light on and off. You know the one I mean. It must have a five-hundred watt bulb fitted, it's so bright. It's the one over their front porch. It looks like a search light on a prison wall, when they're flashing it on and off. Like a warning saying "we're watching you!" They must keep watch for hours, just waiting for us to return, and they never fail to pull the light trick. They do it to Brian and Brenda too. Hasn't it happened to you?' asked Alex.

'No,' mused Gordon, 'but then again I don't think we've been out after dark yet. Maybe our turn will come!'

The conversation carried on in that vein for most of the evening with one person, or another, asking whether or not they had such and such a trick played on them. *Oh boy, did we have crazy neighbours.*

As we left the party, long after midnight, our brains were buzzing with the information we'd been bombarded with. 'Blimey! I don't know what to make of all that do you?' I asked Gordon.

'Mmmh, definitely an interesting evening. I'm intrigued about the Luscombes I must say, and can't wait to meet them all. It looks as if those good old boys will have to take very good care of their sister, if they want to stay around here,' he replied.

I could imagine that having to leave this place would be more than a wrench for them. The moon was flooding the path with iridescent light and, above us, was a blanket of twinkling stars in the clear black sky. The smell on the air was sweet from unseen wild herbs and flowers. I would be very sorry to have to leave this place myself and it was all relatively new to me. We'd had a very good evening; our first social occasion for sometime but we needed to get some sleep before work started with its usual relentless regularity the next day.

Our next major job was to clear out the far section of the Atcost building, where the bullocks had been. On investigating, we could see that the dung was much too deep and compacted, to shovel by hand. The whole of that next week saw Gordon with a hired mini digger, monotonously clearing it all out. When the load in our tipper trailer was piled high, he would then drive it up into the orchard, with the Land Rover. He did his best to scatter it as efficiently as he could, but this was no easy task, as the clumps of muck were almost welded together.

It rained continuously the whole of that week. Our booted feet got stuck in the mud. The dogs got stuck in the mud. Eventually, even the trusty old Land Rover got stuck in the mud. The only winner was the dung, which got well and truly watered in. At least the trees and grass were going to benefit from this marathon task. We should get a bumper crop of apples this year. We just had to hope it wouldn't affect the taste!

After the digger had completely cleared the floor of the end section, I set to with the water hose and a hard broom. I felt very grateful that it was still only spring. At least the flies and heat weren't bothering me. The smell was another thing entirely. Water and manure do have an unfortunate resultant aroma.

By the end of the week the job was accomplished. The whole barn looked a completely different place, almost scrubbed clean. The only casualty had been the mini-digger. Gordon had got a bit slaphappy with it and crashed the side of the cab into the wall of the building, resulting in a broken side window. *Whoops!*

I hadn't, until this point, realised the sheer size of this agricultural building. It was absolutely cavernous. It must have been nearly a hundred feet long by probably forty-five feet wide and thirty feet high. It was divided into three sections by short block walls topped by metal railings, one of which now held the caravan at one end and the bales of hay at the other. The remaining two sections were now empty. Well, they were most of the time! The hens and cocks still roosted at night on the rails at the far end. We had tried to discourage them but, as the barn doors were only half doors and our feathered friends' wings had not been clipped, there was little we could do to stop them flying in.

We felt incredibly happy and content to stand in the Atcost barn and see the results of our labours. If that magical day should ever arrive, when we moved out of the caravan and into our newly converted barn, then this Atcost building would make a magnificent space to store just about anything you could mention.

Another bonus, at the end of that week, had been an unexpected visitor. A young man in tight, snazzy black leather gear drove in, on a very smart looking motorcycle. He came with a message from his father who, apparently, owned both the caravan and the trailer. The lad's father sent his apologies, but was unable to move his possessions at the present moment as he was incapacitated, with a leg in plaster. As soon as his broken leg had mended, he would come and introduce himself and remove his possessions. As a token of his appreciation for our fortitude in storing his property, he sent us a bottle of very nice wine. *Cheers!*

CHAPTER THREE
FEATHERS AND FUR AND JUST A LITTLE WATER

It had been a sleepless night - thanks to the hens. Or should I say cockerels? It was difficult to know just how many hens and cocks we actually possessed but, there's one thing for sure, there were definitely more of the male variety than was healthy. The hen/cock ratio in the flock appeared to fluctuate; probably thanks to the odd fox in the vicinity. With unclipped wings, they had the freedom of the farm and had always roosted at night in the Atcost barn, where we now had the caravan parked. They were there first, after all.

The cocks had fabulously coloured plumage, running the whole spectrum of green, down to an iridescent black. There was one extremely handsome chap, mainly Sherwood Forest green, with snow-white streaks on the edge of his feathers. He would strut importantly around the yard all day, with an entourage of female admirers in his wake. The cocks did seem to outnumber the hens though, by at least five to one, so it was no wonder the girls were always going broody. The flock nested wherever the fancy took them, but favourite spots seemed to be on any little ledge, wherever a plentiful supply of old dry straw had gathered over the years. There was always a good pile of it blown against the partition walls in the Atcost building.

I always tried to collect the eggs as soon as I heard that certain squawking that, I now knew, denoted the production of a steaming egg. I had thought at first, in my ignorance, that someone was murdering them but, later, got to know it was a triumphant egg call. The Atcost building afforded them a large area in which to lay their eggs, some very craftily hidden. It wasn't always convenient, if I was in the middle of a job, to seek their hidey-holes out and collect them. If left to their own devices, the hens would take up residence and sit like glue on their little clutch of eggs; which nothing in the world would persuade them to give up.

The latest hen to refuse to get off her little nest, had been swiftly followed by two others, and there had been no end of squabbles amongst them. Every time one of them got off her patch for a call of nature, or something to eat, one of the others would quickly take her place and then refuse to move, when the rightful owner came back to reclaim her clutch of eggs. There was also the strange phenomenon of egg nabbing going on. They would actually go and steal an egg from another hen's nest by rolling it along the floor and into their own chosen spot, with just a beak to guide them. Rather like a poultry version of shove ha'penny. At one point, there were at least ten eggs in one nest and only one in each of the others: some hens were obviously better at the old egg nabbing lark than others. It was more the behaviour you would expect from a kindergarten class of two-year-olds than a flock of hens.

When the first batch of chicks broke through their shells, there was one almighty row. All three hens tried to claim the honour. I must admit it was difficult to tell exactly whose eggs were whose by that time, anyway. Some of the chicks were definitely starting to suffer, thanks to the amount of bickering going on. The poor little mites were no sooner clear of all the debris of birth, when they became subjects of a battle royal. First one hen would grab it by the scruff of its neck, or wing, only to be quickly chased by another hen-in-combat, ready to fight to the death for "her" chick. I couldn't really see how any intervention on my part would help the situation and had little choice but to leave them to it. Only the strong survive, so they say. All this was going on in a corner just by the side of the caravan. Total bedlam.

The first time I was roused from a deep sleep, there was a cacophony of noise coming from outside. Squawking, screeching and lots of cockadoodle-do-ing. *What on earth was happening?* I shot out of bed, fumbling in the dark for my shoes. The noise was even worse when I opened the door; shining my torch in an arc, backwards and forwards, trying to find the cause. I could see a speckled hen trying to round up three of her chicks, which were madly running around in all directions. Then I spotted the problem: I could see the red reflective eyes of a cat, crouched not far from the chicks, just making ready to pounce. I could see one broken little body by the side of the cat, it looked very much as if, unfortunately, the cat had already nabbed one of the brood and had come back for some more. When the cat saw that I meant business with a broom in my hand, it beat a hasty retreat, with its prey clasped firmly in its jaws.

That then left me with the three chicks to round up and put back into the nest. When they were first hatched I'd propped a plank of wood against the ledge where the nest was - about two feet off the ground - so that, if they'd had a mind to, the chicks could get up and down on their own. I couldn't imagine how they were going to manage to scale the wall, without my device. I hadn't actually seen any of them using this innovative staircase before and, even in extremis, they definitely seemed disinclined to do so. My rounding up technique consisted of chasing madly after each chick, trying to corner it, quickly scooping it into my two hands and then giving it a bit of a heave up into its nest. Easier said than done, believe me, especially in the dark. The torch kept getting in the way, as I needed both hands free to scoop and heave. In the end I resorted to transferring it to my mouth at the last minute.

After successfully restoring the brood to their nest, with mum clucking anxiously around them, it was back to bed for me. The clock told me it was only two am. I snuggled up against Gordon's back trying to get warm again.

The second disturbance came not long after I had collapsed back into bed. Again I flew out of the caravan, frantically shining the torch around, hoping that it wasn't the cat after the chicks again. No, good, they were all OK. I could see them all snuggled in the straw. So what was the noise all about? That was when I spotted a bundle of something that shouldn't have been there. It was over on the far side of the barn and, in the gloom, I couldn't quite make out exactly what it was. Ducking under the rails, I drew close enough to see that it was a kitten lying in the middle of the floor. It still had its eyes closed, and couldn't have been many days old, but was managing to make enough noise for something so small. Presumably, the cat after the chicks had been moving her brood and, hopefully, only temporarily abandoned this kitten. What on earth was I going to do if she didn't retrieve it? It was a freezing cold night and the kitten was in a very exposed position, near one of the draughty doors. I decided that if I tried to cover the kitten, or move it, the cat would more than likely smell my interference, and there was a chance that she would just reject the kitten completely. There seemed nothing for it but to go back to bed and hope that the mother would return soon. Once more I crawled into bed and cuddled up against Gordon's warm body.

Loud squawks and screeches. Another racket! *What now?* This was certainly getting monotonous. Again the torch scanned round the barn. The kitten had gone, so maybe it had been the cat returning, to collect its offspring, that had caused this disturbance. I could only hope it was as simple as that. In the dim light of the torch, I could see that there were three cocks roosting on a metal rail, just in front of an old cattle drinking water trough, not far from where I was standing. This seemed to be where the noise was coming from. Trying not to trip over anything in the dark, I shone the torch onto them. *Nothing.* The torch batteries were starting to go and the light was now nothing but a feeble orange glow only lighting up the dust motes in the air. Moving the torch around a little, I could just see three yellow chicks on the surface of the black, oily, water. One looked very dead – part floating, part submerged - but the other two were actually swimming. *Swimming!* They were flapping their wings up and down and appeared to be keeping themselves afloat. What on earth did they think they were doing, going for a dip at this time of night? I quickly scooped up one chick, depositing it in its nest, then the other. They both stood there shaking themselves, probably not best pleased at having their fun terminated. Unfortunately the one that looked dead - was. During the night that made a total of two chicks lost. At this rate there wouldn't be too many of them managing to grow into hens – or, just my luck - cocks.

Climbing wearily back into bed, I was thankfully left in peace, for what remained of that night.

Gordon was already dressed and out of the caravan before six that morning. Although he had stayed firmly attached to his mattress during the commotion the night before, I think he had had enough of the restless night, and decided to make an early start. He was grumbling a bit about being forced out of bed. He was normally an early riser, always preferring to start the day by taking his first cup of tea for a stroll round the site. He liked to see what we had achieved so far and plan what needed doing next. In the early mornings, just after sun up, his thought processes were at their best and he enjoyed the solitude before the rest of the world joined him.

We had primarily been clearing the site up to now, not wanting to commit ourselves too heavily financially before that vital link was obtained – planning permission. This strategy, although infinitely more sensible, had proved a little too soul destroying for our liking. We had decided to have the scaffolding erected, in order that Gordon could take a closer look at the work that would need to be done to the roof. Since its erection, he had spent some time on the structure, deciding how best to tackle the mammoth task of stripping the roof, without exposing the cob walls too much

We still didn't have access to the whole of the inside of the barn yet, thanks to the pack of dogs still in the top half. The ewe and her lamb had braved the cold and gone to join the rest of the flock in the meadow. They had only needed the sanctuary of the barn for a few days before the lamb was robust enough for normal life. Their departure had given us access to the lower section of the barn at long last, and we had now pretty well managed to clear it of the trappings of its agricultural past. The old milking stalls, with their ironwork railings and head collars, had been gouged out and had now gone to the scrap merchant's. The old concrete floor – on quite a few different levels to originally allow for drainage – had been jackhammered up and removed. The final digging out for the new floor slab was waiting to be done, if and when, we finally got access to the whole building.

With his cup of tea still in his hand, Gordon climbed the scaffolding attached to the front of the barn, to inspect the work that had been done the previous day. He was musing and pondering, as he was prone to do, lying flat on the scaffold and looking up at the roof. He was planning how he would attempt to lift the old roof off and replace it with a new covering. He heard the back door of the farmhouse bang. Looking down, he saw Ted, as usual in his navy greatcoat and wellies, clutching what looked like a rusty biscuit tin full of water. After strolling around the yard, Ted put the tin on the top of the wall, by the gate, and proceeded to give his face and ears a perfunctory wash. After his ablutions were done, he dried himself on a small frayed square of dirty grey towel that he'd pulled from his pocket. He then chucked the water onto the roses and went back inside the house. Well - that solved one puzzle anyway - he did wash occasionally, even if it was only one very small part of him!

Later that morning I encountered the rest of our mysterious neighbours from the farmhouse, just as they were driving off in their car. We knew that there were only now three brothers and one sister living in the house. We'd seen all three of the brothers from time to time but only from a distance. We had so far only managed to actually speak to Ted, despite the fact we'd been living on site for almost two months. For some odd reason we got the impression that they were avoiding us.

The man was standing holding the gate to the farmhouse yard, waiting for the small white car to pass through.

''Ow do,' he said, raising his hat.

'Hi, how are you?' I replied.

'Me eyes is awful missis,' he said as he spat a gob of saliva accurately to his left, barely missing his boot.

He was small in stature and looked to be in his seventies. He had a deeply creased face that looked thoroughly lived in, with a pugilist's knobbly nose. He had a cheery, friendly grin on his face and I could see the likeness to Ted, although this brother was older and had a lean, wiry physique. He was dressed in a shabby suit and waistcoat, made of thick brown checked cloth, a rose bud in his buttonhole. His shirt above his waistcoat was grey and grubby looking, with a threadbare collar encircled by a liberally stained tie.

'I be 'earing all about you lot from our Ted. I'm George and that there is our James and Verity.'

As he spoke, the white car at last crashed into gear and moved forward through the gate, jerking to a stop just a few yards past us. The man in the driver's seat waved through the window and shouted, 'we'em off for a bit of the old shoppin' can we get ye ought?'

'No ... no thanks, that's very kind of you but I'm going into town later myself.' As I was talking, I walked towards the car, hoping against hope, to get a better look at the other two elusive members of the Luscombe family. James held his hand out of the window and shook mine. He was perhaps slightly younger than George, but looked very similar in build. He, like his brother, was wearing a thick brown tweed suit with a jaunty hat on his head.

Unfortunately I was only able to get a fleeting glimpse of Verity as, after James had shaken my hand, he immediately threw the car into gear and shot up the lane. I just got a brief impression of a very diminutive lady, with a tartan rug wrapped round her legs. She almost looked embalmed: she was heavily made up, the rouge on her extremely white skin making her cheeks stand out like round rosy apples. I had often seen her shadowy figure looking through the grimy window of one of the rooms downstairs but never seen her outside before. She unquestionably looked pale enough never to have been out of doors in her life.

They certainly seemed like jovial neighbours, or at least they were when they went out. Let's hope it would last and that Brian's experiences didn't rub off on us.

Not long after the Luscombes had left for town, a police car drove through the gate. The uniformed officer greeted us and we shook hands. 'I've just come along to introduce myself. I'm your community policeman. Just thought I'd come and give you a bit of friendly advice about your neighbours,' he said.

'Oh, yes?' queried Gordon, frowning slightly.

'The Luscombes can be a bit of a problem: they're well known to us for causing trouble in the neighbourhood. It's a good idea to tread a little bit softly with them and try and keep out of their way. While they're on the drink, they're more than likely to throw a punch or two,' said the officer. Even more alarmingly, he continued, 'They've had most of their firearms confiscated, so you wont have that problem to contend with, the way some of your neighbours have in the past.'

'We've been warned already, but thanks for the tip. I can see that they're used to having their own way around here and don't brook any interference from outsiders with good grace,' said Gordon.

'No, you're right there. They've been King of the Castle for too long and can't seem to forget it.'

'*Most* of their firearms, did you say?' queried Gordon, his voice raising an octave or two, suddenly picking up on the fact that there might still be guns in the farmhouse.

'Well,' the officer reflected, 'I think they might still have some old hunting rifle or other but that's about all. After the last incidence, they had their firearms licence revoked and had to hand all their guns in. You never quite know though, with people like them, what they've got stashed away in the attics'.

'You surely don't think there's still any likelihood of them being dangerous do you?'

'No...no I'm sure they're finished with all that now. I think they've learned their lessons the hard way. They've been quiet enough for the past few months, at least. As I said, just avoid them when they've been out drinking and you should have no worries. Well, just you take it easy now and things will be fine, I'm sure.' He gave a friendly wave and was on his way leaving us with a plenty of food for thought.

Unfortunately, the Luscombes' goodwill didn't seem to have lasted from our morning chat. I was unlucky enough to be still working in the lane when they got back. It must have been about three-thirty by that time, and I could see that George looked very unsteady on his feet. He almost staggered down the lane, after allowed the car to drive through the top gate. The car sedately passed by me and, although I waved and smiled, both occupants looked straight ahead.

Taking me completely by surprise, I hadn't noticed George approaching. 'Load of bleddy mushrooms you lot are,' he almost yelled in my face, spittle and froth gathering at the sides of his mouth.

'Wa...what are you going on about?' I babbled, shocked at the sudden verbal attack.

'Come up out o' no where!' he shouted gleefully.

I was a good bit taken back by this onslaught: it took me a couple of seconds but I eventually managed to smartly chalk up, 'Well, at least we are still going up!'

He almost went purple in the face at that remark and I thought it the better part of valour to high tail it out of there. He stank of spirits and pretty obviously wasn't going to talk any sense.

As I hurriedly made my way back down the lane, towards the sanctuary of our own gate, I could hear him still ranting

'Load of rubbish. Stealing a man's property. You bleddy gypsies is all the same. You won't be here much longer, just mark my words.' Thankfully, I managed to find myself a task on the other side of the site for what was left of the working day and didn't see any more of my obnoxious neighbours.

In the days that followed, I became very proficient at finding a job well away from the lane, and used to be able to time their homecomings almost exactly. As long as I was out of sight and sound when they returned from their boozing session, then I was free from their insults, deceiving myself that all was well between us.

CHAPTER FOUR
AN ASSIGNATION IN THE DAIRY

Ted we used to meet on a daily basis. He always seemed to be hanging over our gate, mooching around, just waiting for one of us to appear. I suppose this, to some extent, was a step in the right direction, as he was at least outside the gate waiting to be invited in. Mind you, the shiny new padlock, attached to the bolt of the gate, might have indicated to him that his uninvited presence on our property was no longer required. Certainly our verbal requests had fallen on stony ground. Our daily goings on appeared to be an endless source of wonder to him. We were obviously the biggest source of entertainment he'd had since electricity had come to the farm.

In many ways he was a likeable old character: much more predictable than his two brothers. One reason for this may have been that he always stayed at home when the others went into town and, therefore, didn't get involved in their heavy drinking sessions. His tipple was cider, which he liked to carry around in the pocket of his greatcoat, while he applied himself to the odd jobs around the farm.

He was a simple soul and holding a conversation with him was nothing but a profound strain. His Devonian accent was uncommonly thick, sometimes defying interpretation. I became more adept at translation as time went on, quite enjoying our little chats, and even managing to understand most of what he said to me! He had a wealth of knowledge and experiences that he relished passing on.

Like a lot of old people, the past was more interesting to him than the present. He loved yarning and liked to tell me about the old days when his parents were still alive. He'd been telling me a tale about how his family used to brew cider with the apples from the orchard that now belonged to us. The old cider press, which was inside the cob barn, was used every year to produce an extremely potent brew. Their cider, apparently, was renowned locally for its taste and strength, selling like hot cakes at the farm gate.

Anxious that morning to keep me talking, he told me he'd show me where it all happened. Glad of any opportunity to see inside the section of the barn that was out of bounds, I followed him round to the side door. With the yelping dogs firmly secured behind a door leading to the middle sector, he took me inside the top portion to show me where the apples had been turned into cider. While we were in there, he pointed out what looked like, to me, some sort of box on legs, cobbled together from very rough-hewn wood. He told me that this was where he had once kept his ferrets, but now used to store last year's crop of "keeping" apples. He told me that the varieties of these apples were very old and rare - some sharp, some sweet - just the perfect mix for cider. He thought it was possible that some of these varieties had now vanished from most modern orchards.

The apples on the trees outside were only just starting to fill out: it wasn't so very long ago that the whole orchard had been covered in blossom. My slow brain easily worked out that those in the ex-ferret box must have been there for almost a year. He lifted the lid and out drifted a faint musty aroma of apples; almost a cinnamony smell. He carefully peeled the newspaper from one large orb and bit into the side of it. It looked perfectly OK to me and he said, with the juice still running down his chin and onto his chest, that they tasted 'bootiful'. He unwrapped one or two others to show me the different varieties he'd stored. There were one or two that were bruised, or turning mouldy, but the majority of them still seemed in near perfect condition. After almost a year of storage that wasn't too bad at all.

After Ted had wandered off, I went to tell Gordon of his latest anecdote. I wanted to show him the different varieties of fruit our trees would be producing but, disappointingly, he was busy and couldn't leave what he was doing immediately. Later, when he was free, I took him to see my find. Gingerly opening the barn door in case the dogs were back inside the top section, I went inside. The dogs, thank goodness were nowhere to be seen or heard. Neither was the ferret box or its contents. They had disappeared. *Did I imagine the whole thing?*

'Well, where is this ferret box you've been on about?' asked Gordon

'It was here! I know it's a bit dim in here but it definitely was over in that corner. It can't go missing just like that, surely?' I asked reasonably.

'Bloody Grimms again, I suppose. They'll have nicked it just like that milk churn you were saving to put outside the front door,' said Gordon in disgust.

We had unearthed one or two old farm implements, buried among the rubble, including a lovely old milk churn that I had spent many hours lovingly cleaning so that, when I eventually had a front door, it could sit outside. We'd left the churn and the rest of the bits and pieces in a corner of the Atcost barn, all cleaned up and ready to give back a little of the old character to the barn, when the conversion was completed. One day after we had both been into town, the milk churn went missing. Another one of those mysteries we'd probably never get to the bottom of, but we had a sneaky suspicion that one of the Luscombes had been having a look around and decided that the cleaned up churn looked too good to miss.

We had begun to call the Luscombes "Grimms" after I had seen two of the brothers at dimpsie. Dimpsie is Devon speak for the time just before full dark when the light is just fading - twilight I supposed. They were making their way across a field: it was drizzling and the light was getting pretty shadowy when I saw them, crouched against the skyline. They had sacks covering their heads, as some kind of protection from the steady downpour. The sacks were slit up one seam, leaving an effective envelope to cover their heads and shoulders. They could have been the inspiration for an illustration in a fairy tale, as they looked like two Victorian villains making their way back after body snatching. Perhaps I should have christened them Burke and Hare but, at the time, Grimms seemed more appropriate and the nickname firmly stuck.

Ted's accent had always been a bit of a struggle for me. On occasion, he could almost have been speaking a foreign language. When we conversed, I would often as not have to ask him to repeat a sentence more than once. Very occasionally, if I still couldn't grasp what he was trying to say to me, I would take the cowards way out – just nod and smile and give a 'mmhmm' every so often as an encouragement. It didn't seem to matter; he seemed happy enough to carry on our conversations and my lack of response didn't, apparently, faze him at all.

One day I met my Waterloo and he stumped me completely. He saw me in the lane and asked, "Ave ee seen them there glinnies abouts?'

Glinnies?....glinnies? Did I hear him correctly? 'What have you lost Ted?'

'Them there glinnies. I aint seem 'em for days now, and I'm gettin' a mite worried.'

'Glinnies'? I queried, hoping for further enlightenment.

'Yes...glinnies...'ave you seen 'em abouts?' He looked at me a bit quizzically; seemingly unable to make up his mind whether I was hard of hearing or just plain stupid.

It was obviously time for me to come clean. I hadn't a clue what he was talking about and it didn't seem from the turn the conversation was taking that I ever would.

'Sorry Ted, I really don't know what a glinnie is,' I humbly confessed.

'You don't know what a glinnie is, maid? Well I'm buggered!'

Still none the wiser, I patiently waited for him to enlighten me.

'Well, it's... it's one of them there humpity little fowl like things.'

'Fowl like things? What? Chickens do you mean?'

'Nooo! O' course not!' He started to shake with laughter. 'Them there grey fowl things, I means.'

'Do you mean those things that are grey and white and sound like a goose with a bad case of hiccups?' By this time I thought that I had worked out that he just might possibly mean guinea fowl.

'That *is* right,' he agreed.

'Oh, those things! I know what you mean now. Yes, I saw them yesterday on our bit of land, on the other side of the stream. I think one of them is sitting on eggs. She flew into a bit of a rage when I got too close to her, while I was walking the dogs. Trouble is that she's on the ground, just behind that blackthorn bush, and probably not very well hidden from the foxes.'

'I'll go and see. Maybees I can persuade 'er to move somewhere safer.' And off he went, taking a couple of the muddy-brown dogs with him for company.

Every weekend, usually on Friday or Saturday evening, we started to notice that Ted was furtively hanging around the lane. He seemed to do this on such a regular basis, that we were intrigued, despite ourselves, and took to watching him. We found out, eventually, that he was meeting someone. It looked like a secret assignation, as there was a bit of sneaking around going on, and we were more intrigued than ever. A man would park his car on the main road and then walk down the lane to where Ted would be waiting for him. They would then spend half-an-hour or so deep in conversation and then the man would disappear back up the lane to his car.

These meetings seemed to creep into our conversations in the weeks that followed. We did, of course, eventually find the solution to the enigma. One evening when Ted was late for the tryst, we bumped into the mystery man. No, we hadn't uncovered an illicit spying ring: he was brother Vincent!

We found him hanging over the gate taking a look at some of the work we had been doing, while he waited for Ted to arrive. He was as large as Ted, both tall and rotund, and was smartly dressed, in an immaculately clean shirt complete with natty tie. His navy trousers had crisp creases down the front: his black shoes meticulously polished. When he told us that he was Ted's younger brother, we were incredulous. He actually looked normal! And clean! And he didn't smell of anything other than aftershave! He was also a reasonably articulate man, without the broad accent of his brothers. You could have knocked us down with a feather.

While Vince was waiting for Ted to appear, he told us his part in the Luscombe family saga. Years ago, when he'd first left home to get married, he'd asked for his share of the farm, so that he could start up a business of his own. A reasonable thing to request, according to him, but a view obviously not shared by his brothers. The valuation of the farm was set at a high figure, and coming up with Vince's share had caused many problems and disagreements. Not long after Vince had been paid out, the market value of the farm dropped and George and James, apparently, felt cheated, refusing to ever see him again. They saw Vince as the harbinger of all their monetary problems and, in some obscure way, also responsible for the gambling debts and the bank foreclosing on them.

Evidently Ted didn't seem to share the feelings of his two brothers, or hold any grudges. As Vince wasn't welcome inside the house, whenever he came to call on Ted they had to meet outside, or in one of the barns. Now that we had bought the barns, they had nowhere to meet when it rained. Or did they? Most of the time when it was raining, or when the weather was unpleasant, we were inside ourselves so hadn't noticed their new rendezvous – the dairy. Vince told us that as the weather had turned colder, they had now started to use the dairy. He was kind enough to ask our permission to carry on using it. We were glad to be able to give that permission - well at least until such time as the building was no longer a dairy and more of an entrance hall. Then they may have difficulties!

Our first real yarn with Vince's eldest brother, George, came when we had started to erect a post-and-rail fence all the way from the gate, leading from the main road, down to the barns. We had been in a bit of a quandary as to what to do with our land. It wasn't a great deal – only about three acres - laid to grass, but it had been getting a little out of hand, with the grass falling in on itself, thanks to the volume of rain we'd had lately and the warmer weather. We'd talked about having a donkey, but found out from the Donkey Sanctuary that they were companionable animals, which would have meant having two or more. Goats were out. Most definitely out! We had already seen that they eat literally everything - including the trees. They had even managed to eat a great deal of the embryo apples, by standing on their hind legs and leaning against the trunks. Fickle animals that they were, they didn't seem remotely interested in the lush grass that needed eating.

After making a few enquiries locally, we'd met a lovely lady called Belle, who was Henry's mother. We'd met Henry quite a few times in the pub. He was the local heartthrob, both young and handsome, and was much admired by all the girls. After discussing our problems over a few pints, he said he thought his mother might be interested in helping us out. Belle liked to keep a few sheep but didn't have too much grass-keep of her own. She certainly didn't allow the grass to grow under her feet, arriving next day to look around at the meadow and we managed to strike an excellent bargain. She would bring her sheep to eat our grass and then, when she was having an animal slaughtered, would pay us with some lamb for the table. A perfect arrangement. We needed to erect some kind of fencing, however, and thought we might as well do a "proper job" and make it a post-and-rail.

We had started up by the main entrance; concreting the first of the posts into the ground, and had got five of them firmly established, when a figure came up the lane. He was dressed in the usual uniform adopted by the Grimms, an old trench coat, tied round the middle with bright orange binder twine; a greasy pork pie hat perched on his head. He looked a little like a scarecrow, but not quite so well dressed. He was carrying a thick staff that had once been the limb of a tree and, as he walked, he liberally dug this into the surface of the lane. Two slinking curs closely followed him.

''Ow do,' he said as soon as he was within hailing distance.

'Good afternoon,' casually replied Gordon, carrying on with hammering in the next post.

'Got somethin' you might be int'rested in young man,' George said, and with that, he spat a great gob of saliva expertly at the foot of the post.

Outwardly unperturbed Gordon said 'Oh, yes... and what might that be?'

'That there bit of land you're standing on. It might be for sale - at the right price,' he said, triumphantly. I could smell his breath, now that he was close-to, and it was more than generously laced with alcohol. I could also detect that unmistakable Grimm stench that certainly didn't come from a bottle and had taken many months to accumulate. It reminded me of the bouquet released when we unearthed those sheep recently.

This conversation was taking a strange turn. The question about the land was a puzzle, because Gordon and I were actually standing on our own piece of ground. We had bought all the land down from the main road, including the lane itself, in all about three acres, from the Receiver.

'What bit of land would that be, then?' enquired Gordon, hoping for some enlightenment.

'Why this piece and that little bittie bit this side of the brook,' he said gesturing towards where we were standing, and beyond, to the main road.

'What would you say was a fair price then?' asked Gordon, obviously deciding on a "wait and see" strategy.

'Well us ud 'ave to 'ave a little chit chat about that over a glass or two, I'm thinking, what do you say about that?' he said, with relish. Another wad of saliva showered down.

'I'm too busy right now for drinking, just tell me what you think it's worth and why I should buy it,' demanded Gordon.

'Well...it'd go just proper with the little bit you got right now round the barns. You could keep a few sheep there, if you wanted.'

'Yes, I know I could keep a few sheep – that's why I'm fencing it. Why would I be fencing anyone else's land but my own, do you think?' Gordon glanced up at the sky, no doubt looking for divine intervention.

'Well, I dunno, do I? You could jus' be doin' folk a favour - just bein' neighbourly like,' replied George, shuffling backwards and forwards making himself comfortable on his staff, preparing himself for a long conversation.

'Look...why don't you stop wasting my time?' Gordon said, getting tired of the game, his exasperation beginning to show itself. 'You know I already own this bit of land. What you obviously don't know, is that I know it too. What kind of a fool do you take me for?'

'No, you'm wrong there. That bit is our'n all right.' George had started to shout by now. 'And this lane is our'n too and you'm trespassing.'

'No it's not yours and you know it. We bought it at auction from the bank.'

'It's our'n I tells you.' George reiterated, his voice now raising a few notches in pitch.

'Oh come on! If you hadn't lost it, with your drinking and gambling, we wouldn't be here at all. You should be ashamed of yourselves: grown men acting in that fashion, loosing your inheritance,' Gordon was shouting himself by now.

After hearing the last few of Gordon's words, George's face darkened with temper: he was grinding his porcelain teeth and muttering under his whisky soaked breath.

'You'm big liars, you gypsies. I got papers I 'ave, I can prove I owns it,' he spluttered.

'Funny you should say that, I've got papers too, that prove *I* own it.' Gordon, looking unconcerned, carried on with his hammering.

Realising that he had met his match, and was not going to pull the wool over Gordon's eyes, George called his dogs and went off to where his sheep were grazing.

A couple of hours later we were still concreting in posts, when he sauntered up the lane again. He was grinning and appeared to have recovered his lost temper.

'Are you still at it then? You works too 'ard you do. Need to take more time to 'ave a chitchat and a drink with friends.' *So that was what he was after...*

'Well there's lots to do, if you want to lend us a hand,' was Gordon's response.

'Oooh, I would boyee, if on'y I was ten years younger, but me eyes is awful now,' came the excuse. ''Ave you thought about our little chat back along?' he continued.

'What would that be about?' asked Gordon, tongue-in-cheek.

'About that little bit of land you might like to buy off me?' the cheeky old devil replied.

'I'll give it to you George, you've got bottle! You're a swindling old devil. You know, and I know, who owns this bit of land, so just let it drop will you? I've nearly finished here anyway, come back to the caravan with me and we'll find a bottle or two to open. What do you say?'

'Sounds a proper job to me boyee!' George reiterated his agreement with another huge gob of saliva, which ricocheted off the ground near my feet.

Peace was bought with a glass or two of brandy.

CHAPTER FIVE
THE HEN HOUSE DEBACLE

We had been living on site for a couple of months by now.
The pattern of our lives, in our new home, was becoming
established. Just the same as the pattern of our neighbours'
lives became apparent.

They were early risers, like most country folk. Unless we
were out of the caravan well before six in the morning, we
didn't catch sight of Ted on his usual patrols of our property.
As the mornings became lighter and warmer, we both felt
the need to get up at dawn, just for the sheer joy of seeing
the sunrise. Even if we did have a lie in, it didn't take a
detective to work out that Ted had been around. One big
give-away, was the corn scattered all over the newly swept
yard.

We had seen quite a few rats when we'd first arrived -
both dead and alive - and were anxious to discourage these
as much as possible. With so much corn left lying around,
they were, not surprisingly, proving difficult to control. We
were feeding the hens ourselves by this time, in any case, so
couldn't see why Ted should feel the need to feed them as
well. They were daily getting a double ration, and most of it
was left lying around on the ground at the end of the day,
unwanted except by the rats.

Good old Sharp may have been getting on in years, but he
proved himself an excellent ratter. It was quite unnerving to

walk back through the Atcost with him by my side and suddenly see him dart out and grab a rat by the neck. He would give it a thorough shake and then drop it at my feet. He had obviously been well trained at some point in his career. The rats he caught might well have ingested poison; making an uncomfortable meal for him, had he been tempted to eat them.

Gordon had asked Ted on more than one occasion not to feed the hens on our patch. More importantly, as far as we were concerned, was the fact that he was still wandering around on our property - uninvited and unwelcome. We both disliked finding him poking and prying into odd corners, and didn't want to encourage Ted to think he could still roam round whenever he wanted to. After being interrupted over a meal, when he'd just come and sat down at the table with us, we'd resorted to padlocking the entrance gate in the hope that this would discourage him. He didn't give the impression of being the most athletic of chaps nevertheless he still seemed to find his way over the fence. It was all very well at the moment when it was still a building site, but our imagination had taken us beyond this, to when we would perhaps be sunbathing in the garden, only to be disturbed by Ted shuffling by.

Gordon had asked him to leave the feeding of the hens entirely to us or, better still, to scatter the corn in the lane or on their own land. We had hoped that this might even encourage the hens to move their sleeping quarters permanently and take up residence in the farmyard. They were, after all, the Luscombes' hens, not ours. Although I enjoyed seeing them around, I could well do without them roosting in our barn next to the caravan and disturbing our very precious sleep.

Ted had obviously ignored the heavy hints as, for the last few mornings, we'd noticed him with his bucket of corn, scattering it willy-nilly on the concrete apron in front of the Atcost building. He must have, presumably, climbed over the gate to gain entrance to the yard. Gordon caught him again, throwing the corn near the cob barn. 'Ted!' Gordon shouted. 'I thought I'd asked you not to throw that corn there? Could you please either scatter it on your own land or, if not, do it in the lane, there's a good chap. Those rats are going to take us over soon, if you keep leaving that corn here.'

The reply was unintelligible and all he heard was 'mutter mutter' as Ted shuffled off into the distance.

'That's it!' Gordon said as soon as he saw me, absolutely seething with rage. 'I've just about had it up to here with the Luscombes. I've finished with the softly softly approach. It just doesn't work with that lot.'

'What's up now?' I asked.

'Ted just wont take any notice when I ask him not to scatter the corn in the yard. It was seeing him coming out of the cob barn that did it. He'd got a very suspicious bulge in his pocket, I'm sure he's nicked something. Doesn't the man realise we'll be living in that barn soon and that bit of yard will be our garden! He just doesn't seem to get it into his thick head that I own it now, not him and his family! Does he think he's going to go on scattering his corn in our yard forever?' he ranted, well incensed. I didn't think Gordon really expected an answer to that little tirade, so I just ignored him and carried on with my job. The issue as I saw it wasn't the corn in any case, but the fact that Ted was still hanging around us like the sword of Damocles. It was comparable to having a third person living with us, and a very unwelcome third at that!

I felt just as fed up as Gordon did at that particular moment, as I had been allocated the task of carrying the first batch of Thermalite building blocks into the cob barn. They'd been delivered the day before but the lorry, which was a huge articulated truck, had been unable to manoeuvre into the yard and so had deposited them at the entrance. They now needed carrying into the barn, so that they would be ready for the brickies to lay. There were sixteen pallet loads and, as they were pretty heavy, I could only manage to carry two at a time. At this rate I would be doing the job all day. Boring, uninteresting work.

Some of the cob, on the inside of the barn, had crumbled almost to dust over the years. In places there were large gaps, especially around door and window openings, where it had just disappeared. This had more than probably been caused by water penetration. Cob is an ancient but very strong building material, or it is until it gets wet, and then...the walls come tumbling down. Repairs had been carried out over the years, several in blockwork but the majority in red brick. Some of the brick repairs were in very poor condition, mostly old crumbly Victorian bricks, so we had demolished them. The bricklayers were due to start work the next day on the repairs.

'Carry some of those blocks up to the far end of the Atcost building for me will you?' Gordon said.

'What for? I asked.

'I'm fed up with having those hens in the barn with us. I'm going to build them their own quarters...a fabulous new house!' he replied. 'When Ted can't see them, perhaps the silly old bugger will finally get the message and leave us all alone.'

I did as he asked, carrying some of the blocks to where he had set up the cement mixer and water buckets at the side of the Atcost building. Two of the goats came across from their patch on the other side of the stream, to look curiously at what we were doing. Fortunately, Billy - the one with the enormous horns - stayed on the other side. The nanny and kid were welcome to stand and watch as far as I was concerned; nice mild little creatures that they were. Mind you, it was about time they were making tracks to another pasture. They were now, apart from the poultry that is, the only animals we were waiting to get rid of.

The ewe and her lamb had long ago been put out to pasture. The weather had soon become drier and the lamb old enough to cope with the rigours of outdoor life. We had got to know them quite well while they were living beside us in the barn but they were now totally indistinguishable from the other fifty or so sheep in the Luscombes' flock, placidly chomping at the sparse grass in the one remaining field.

Next to go had been the bull. This animal had frightened me to death on more than one occasion when traversing the yard. He was massively built, his short legs looking barely able to take his weight. He didn't move very fast, it's true, but just the sheer thought of entering his domain was enough to give me the willies. I had nagged and nagged at Gordon to get the Luscombes to move the bull, until he totally lost his rag with me. Eventually, of course, the bull had been led away to live in the shippon. He only lasted in his new quarters for less than a week. Ted told us the bull had been sold, as he was now surplus to requirements, or so he said. The lucky farmer, who had purchased him, would be taking him away in a couple of days. Likely tale, we thought. Who on earth would want to buy that old heap? The Luscombes had told us so often that things would be done and never were; needless to say, we were immensely surprised when a lorry arrived to collect him. Obviously there was life in the old bull yet.

I have seen bulls loaded onto lorries before now, and it's definitely easier said than done. While we were doing our last barn conversion, we rented out our grass-keep, which was grazed by a herd of fourteen cows and a bull. When it was time for the bull to be moved, they came mob-handed with six men who erected barriers each side of the loading ramp, but this still didn't stop the bull from charging over the top, crashing through the hedge and escaping up the road. Mayhem! Here there was only the oldish chap driving the lorry and the three Luscombes, to accomplish the loading, and I was quite looking forward to watching this bull escaping up the lane. I needn't have been so contemptuous of their skills, (or was it the bull's placid temperament?) as the task went off without a hitch. I think the bull must have been so bored by his own company that he wanted to leave for pastures new, and was quite willing to co-operate. He very regally and sedately swaggered up the ramp and into the lorry, as if he hadn't a care in the world.

Encouraged by our success in getting rid of the bull, we then tackled our elusive neighbours about the tribe of dogs still in the cob barn. I hadn't been able to see inside most of that barn, thanks to those dogs, and Gordon wanted to make a start clearing all the old farming debris out. We knew that there were all sorts of old tools and heaps of timber in there and wanted it cleared out, ready to start building, as soon as the planning permission came through.

If we needed the Luscombes to do something, then poor Gordon always had the onerous chore of trying to persuade them to play ball. They took absolutely no notice of me whatsoever. He would start off patiently enough, but inevitably end up getting riled and angry with them. Frustration like this he suffered on a daily basis. They seemed to have the art of working round any request to do something, by wrong-footing us. On one hand, we certainly didn't want to antagonise them, as they were our neighbours and, for better or worse, were stuck with each other. They obviously didn't like the fact that their property had gone into our possession without, as far as they were concerned, getting any money out of it. We didn't like the fact that they still acted as if they owned our property and made us feel that we were working for them. They acted like feudal lords who had a couple of serfs to manipulate. Not a very happy situation for us.

The Luscombe family had owned the two barns we had bought at auction for more than fifty years. They originally had the farmhouse, which had a date stone of 1616 inscribed on the front, our barns, plus in excess of one hundred-and-fifty acres of grazing land. When the parents died, the whole estate had been left to the six children. George, James, Ted, and Verity were still living in the house, as none of them had married. There was also another brother, Vince, and sister Kitty, who had both left the farm to get married some years before. The remaining brothers and sister had paid out these two for their share of the farm. Story has it that there was also quite a lot of money left to them, apart from the property, but whether that was just rumour or not we were never sure.

By all accounts, the young Luscombe lads were quite a force to reckon with. They were bad enough now! The locals told us endless tales of the rabble-rousing and carousing that used to go on, when they were all younger, and still living at home. Apparently they were great fighters, and the boys loved nothing more than a good punch up with their neighbours. They'd also spent many nights in the local lock-up, after some fight or other. The Mays, who owned the next farm across the valley, were always having trouble with them. Their son was Rick, who we'd met at Kathleen's house a few weeks before. He was slightly younger than the Luscombes and told us many tales of the bullying and abuse he got from them.

Presumably, the girls had not got into the same kind of trouble as the lads. Verity, especially, must have been left out of this equation, as we'd heard that when she was about five years old, she'd fallen down the stairs. Her injuries were pretty bad, apparently, because she never walked again. Rumour had it that she was not taken to hospital at the time and this was the cause of her paralysis but, I must admit, I did find that very hard to believe.

Later on in life, they took to gambling. They would regularly go to watch the horse racing at Newton Abbot or Exeter. Eventually the gambling debts became so large that they were forced to mortgage the farm, to release some cash. They seemed to be as unsuccessful at gambling, as they were at farming. Apparently the repayments weren't kept up, so the bank had eventually had no alternative but to foreclose and declare them bankrupt. They were now only able to live in the farmhouse as long as Verity was alive. The Luscombes were living on borrowed time, but you'd never have guessed it from their demeanour.

The next surprise in store for us was that the incarcerated pack of dogs disappeared. One day you could hear the dogs barking away when either of us came near to where they were kept, in the top of the cob barn, and the next – total silence. Where they had gone was a mystery, but certainly not a mystery we were anxious to investigate. Not hearing anything for a few days, we very warily crept up the steps at the far end of the barn to listen outside the door. Nothing. Not a sound. Gordon cautiously opened the door with me, very cowardly, hiding behind his back. We expected them to come rushing out at us but no – not a dog in sight.

Gordon had only managed to fleetingly see inside this top end of the barn, when he viewed it before the auction, and then just once again when he took Dave Goodenough, the architect, to have a look before the plans were drawn up. This was my first real look through the whole of the upper section. Although the light filtering in from the open doorway was only very dim, we could just see that the floor was relatively level for about four feet and then sloped gradually down. It didn't look a hard compacted surface but was made up of some sort of fury dry matter. Looking closer we could see it was soil, straw, and guess what – dog muck. Great heaps of grey dusty mummified turds.

Towering in front of us was a monstrous great cider press and, by its side, an apple-feed hopper. The press must have stood ten feet tall with an enormously thick beam on the top. You could just make out that there had been a corresponding beam at the bottom but this was now so rotten, very little of it remained. Both the press and the hopper appeared not to have been touched for years. They were falling to pieces and draped with cobwebs so thick they looked almost like very dirty Nottingham lace. We were definitely going to have our work cut out clearing this lot out.

It was intensely exciting to be able to wander from one end of the barn to the other. Such a luxury to have a free rein at long last. We had been on site for more than two months: only now allowed to see the whole of it. With the architect's drawings in our hands, it was good to look around and see what each part of this crumbling ruin would ultimately become. Our new house was, eventually, going to be far larger than we had at first thought. Of the two sections, we had decided that the top portion was to be left as one enormous space. This would be one storey and would have cathedral ceiling with open exposed beams. The bottom portion would be on two levels. The kitchen, dining room, study and cloakroom would all be downstairs. Upstairs there would be two large bedrooms with en-suites, then three slightly smaller bedrooms and a family sized bathroom. It took us a couple of days for the penny to drop that we had the site to ourselves, but once we had realised, we felt like celebrating. What an achievement!

Our working hours from then on were spent clearing out the mess that had accumulated in the top end of the barn. We dismantled the hopper and chopped up the timbers for eventual burning, on the wood-burning stove, which would be installed in the sitting room. It was all pretty rotten stuff and was fit for nothing, except firewood. The cider press we were forced to leave where it was, as it was all too heavy for just the two of us to cope with. We would definitely need help dismantling that baby. The top beam was absolutely massive: weighing much more than a ton on its own. Unfortunately, this was one of our many mistakes. We thought we had plenty of time to dispose of it. If only we had got rid of it there and then. But more on that later...

We realised very soon that we were going to need a digger to get the floor levels right but didn't want to go to the expense of hiring one until we'd got planning permission. Gordon had already done a certain amount of the work by spade, especially in the areas he knew the digger wouldn't be able to reach, such as round the edges, close to the walls. This was very hard going and after a couple of hours decided he needed a break. Searching around for an easier task to occupy his time until the planning permission arrived, he decided the hens needed sorting out.

To build a new hen house seemed like a good idea at the time. It was a nice change to be actually constructing something instead of knocking it down. It didn't take long for Gordon to get the walls up, and then he put a sloping corrugated iron roof on the top. Inside there was even a rail, half way up the wall, for them to perch on. Gordon added a nice wooden door that had a square shape cut out, almost like a window, complete with wire mesh, so that the occupants would get plenty of fresh air. Fortunately, he did draw the line at curtains! It could have been a little Wendy house. Very smart.

While we were building it, the hens and cocks would be milling up and down the concrete yard but it seemed that as soon as it was finished, and we wanted to introduce them to it, they were nowhere around. The night before, they had disturbed our sleep on a number of occasions, getting very agitated about something or other, and we had a sneaky feeling that one or two of them were missing. Perhaps there was a fox around? Sure enough, just as the sun was going down, we spied a fox scurrying across the field, just this side of the hedge. It was definitely time the flock decided to roost in their new house. With the door shut they would be safe enough from any predator.

Have you ever tried getting hens to roost somewhere other than where they are used to roosting? Not an easy task. We first tried scattering their feed inside, just so they would become accustomed to the feel of it. As soon as the corn was down, one of them would, very cautiously, poke a head round the doorway, scurry in, grab a few beakfuls and then rush out again, as if the hounds of hell were after it. Then another one would go in, and soon a couple of others would follow. But not all of them. The rest would mill around making little disagreeable chuntery sounds. We badly wanted to shut them all inside at night but they had other ideas. We left the door propped open with a brick, just in case any of them should just fancy the new apartment. None of them did. That night they were all to be found roosting on their usual rails inside the Atcost.

Days later the softly softly approach had palled. As soon as we had a few inside, we eventually just shut the door and left them for the night. We reckoned we'd got one quite young hen, a couple of the older ones and one cock. The others meandered around, looking very confused, wondering what was going on, no doubt. They eventually made their way back to their old roosting ground, on the rails just outside the caravan window, nearest to where we attempted to sleep.

Next morning we opened the hen house door to a stampede. Out they all rushed as if they'd been incarcerated for a week. They soon settled down and we scattered some more corn around the outside of the hen house, as well as inside the door, just for good measure. That evening it was no surprise to see that there were fewer takers for the new accommodation. We only managed to get two inside with the corn bait.

Our patience was eventually rewarded, though, as a young black hen, three older speckled ones, and a cock we later named Corky, decided that they would take up permanent residence. I'm not sure whether it was our efforts, or the fact that the fox was still around, that persuaded them it was the safer option.

Unfortunately the fox did see our troop as an easy target, as access to the Atcost at night was a simple enough task for him. Most evenings after we had first spotted him (or her), he fed very well on the Luscombes' poultry. We were glad that we had gone to the effort of building the hen house, at least some of them were saved, and lived very happily in their new abode, rewarding us well with at least a couple of small eggs a day. If the weather was warm, and I was quick enough to collect the eggs before one of the hens decided to go broody, then we would eventually end up with a glut. And what a lovely glut it was: I would spend many a happy hour baking.

The eggs were a total marvel to me, as I had completely forgotten how sensational "real" eggs could be. I was too used to the supermarket variety: it came as a very pleasant surprise to find that our own eggs were almost totally yolk – and what yolk! They almost pulsated with their intense golden sunshine colour. When teatime came around, we would very happily sit munching on a still warm fruitcake, which was almost unnaturally orange in colour. Thanks girls!

CHAPTER SIX
CAUGHT IN THE ACT

We were just congratulating ourselves on our progress so far, when disaster struck, in the form of a visit from the Planning Officer. We were well aware that the barn hadn't got planning permission of any kind when we bought the property, consequently, made it our first priority to get an architect to look at our ideas for the conversion. He had then drawn up some realistic plans and submitted these to the local Planning Department, just a few weeks after the auction. We were now waiting patiently, with fingers crossed, for these to be passed before we started any major work or expense on the main barn; just in case events didn't work as we expected. We knew very well that nothing was going to happen overnight; indeed knew that the whole planning process was a long drawn out procedure. What we didn't know was that until the barns had got planning permission for residential use, we shouldn't have been living there at all! Even in a caravan.

Here was I merrily stringing out a washing line between the apple trees, daily hanging out our laundry to dry: advertising to all and sundry that we were here, living on site, and no mistake. It couldn't have been more obvious could it?

The day the two planning officers arrived, I had a full load on the line: sheets and towels were happily flapping away in the breeze. It was quite early in the morning so they found me in the caravan, making the bed and generally clearing up. *What a give away....*

Gordon had already started work at the top of the orchard, clearing away an old lean-to shelter that had fallen down. I pointed them in his direction and got on with my tasks, totally unaware of the major storm brewing.

Gordon came to find me as soon as they had left and I could see immediately by the scowl on his face that he was not in a good mood.

'What did they say: anything that I ought to know?' I said.

'Huh', he snorted, 'I'll say!'

'Oh, don't tell me it's more problems,' I groaned, 'I don't think I can stand another disaster.' I'd started to get more than a little anxious by this time.

'You can get that washing in for a start! It looks like a bloody gypsy encampment. They've just told me we've got to move the caravan off the site. It looks like we're going to have to find somewhere else to live for a while.'

'You're not serious?' I quavered.

'I'm deadly serious.'

'But why have we got to go?'

'That was the planning officer and her assistant come to tell us that we shouldn't be here at all, not until we've got full residential use. Someone's phoned the Planning Department to inform them that we're living here. I bet it was one of the Luscombes!' he said emphatically, thumping his fist on the table and making the cups jump. 'But...that's not our only problem. We may have to knock the Atcost building down,' Gordon said, glumly. 'Things are not exactly going according to plan, are they?'

He looked utterly tired and despondent. We stood with our arms around each other, trying to give and receive a little comfort. It didn't seem to help much.

'They asked me what our plans were for the Atcost building and I told them what we'd thought of doing. You know...storing a few caravans - maybe a boat or two - even hay - that sort of thing. They said that we definitely wouldn't be able to do any of that, unless we agreed to do some major alterations to the access from the main road. They thought the visibility splay wasn't good enough. That's going to be a major bloody civil engineering job! I can't see us being able to do that, because of the cost. It just wouldn't be worth it. They even said they might make it a condition of the planning consent that we demolish the whole building! I can tell you, that took the wind right out of my sails.' Gordon sat down heavily on the bed, shaking his head in despair.

'The road access is OK; you can see perfectly well to pull out. Why on earth would they want us to alter it?' came my tearful reply.

'God knows! It's worse than that...there were also lots of other niggly things they wanted.'

'Like what?'

'Like not being able to build a cavity wall inside the cob barn. They want us to just lime wash the walls. I mean, we can forget about insulation and damp proofing. They didn't seem in the least concerned when I mentioned them. They also want the cider press left exactly where it is! Right in front of a window! It just doesn't make any sense to me. Why on earth didn't we get rid of it?' He was almost pulling his hair out by the roots in his agitation. 'What a blunder! If we'd trashed it they'd never have even known it existed. I even pointed out the beam at the bottom, where it was completely rotted through, but they still seemed to think it was all worth saving. Even in it's present condition! And they'll probably make us take off all the render on the outside of the cob barn, so that those walls can be lime rendered too. I just don't believe it. It'll take weeks and weeks of work to get all that old render off. It's actually put on over wire mesh, so that will all have to be stripped off before we start.'

Over the years, repairs of sorts had been done to the original sixteenth century barn, and there was now a wire mesh skin completely covering the surface of the cob on the outside. The mesh had then been plastered over, with an extremely strong pebbledash render applied over that. Not particularly attractive but no doubt efficient in keeping the cob dry. The render and plaster would have to be chipped off and the wire mesh removed, before the cob could be lime rendered. The thought of all that time-consuming work was just appalling. Not to mention the extra cost involved.

'Things couldn't look much blacker right now, could they? I suppose I'd better start by getting that washing in and not advertising the fact that we're here, at least,' I said, with a tremor in my voice. I was feeling a little tearful by this time and just wanted to be alone. I set off with the washing basket; the two dogs close at my heels. They seemed to sense that all was not well in our world and were trying to provide comfort where they could.

In the days that followed, we spent many hours discussing what we were to do. We needed to be on site: it was absolutely essential we stayed where we were. The nearest caravan site was more than ten miles away, and the length of time it would take us to travel daily would really eke into our work time. Not only that, but security would be a problem, and it would likely mean one of us having to stay at night to guard the place. There just didn't seem to be any acceptable solution to our problem.

Gordon had already built a large storeroom in the corner of the Atcost, where the hay had once been stored. This was a fairly substantial structure built from four-by-two timber and Stirling board, with a locked door for access. In this enclosure, stacked to the gills, were all our worldly possessions. We had machinery on hire, also stored in the Atcost for safety. There were vast amounts of materials and tools we had bought to start the job, just lying around wherever they had been delivered, or needed. One or two things had already gone missing. They were only small things, admittedly, but we had the feeling that if we weren't there on site, especially at night, then more might go adrift. It was virtually impossible to make the building more secure than it was, thanks to the huge metal doors, and the fact that the structure was designed for animal shelter. It was reasonably easy for someone to climb over the top of the doors; Gordon had done it a few times himself when we had locked ourselves out.

I was also anxious about the fire risk. It had been impossible to insure our furniture, while it was being stored in the barn, and my imagination wasn't letting me get a lot of sleep thinking what some malicious person might achieve with a bit of hay and a match.

We had first moved onto the site in early March: this was now the end of May. Our reasoning had reached the conclusion that we had been here all this time without anyone being the wiser, why not a little longer? They would just have to come and evict us, if necessary. Surely it couldn't be too long until the plans were passed? Would they even be passed at all? Doubts were beginning to creep in. It was nail-biting stuff. We decided to speak to the architect to see if he could hurry things along.

Dave Goodenough, our architect, was a long time friend, and had helped us with the plans on the other barn conversions that we'd done in the past. When we spoke to him, he seemed to think that the plans would be going before the committee at the next planning meeting which, he thought, was in only a few days time. That was enough for us. We were determined to lie low until the decision came through. Our whole future seemed to be hanging in the balance. Would we still be here next week?

Two days after our conversation with Dave, the champagne cork made a spectacular bang into the late afternoon sky. What a relief! We both felt like limp lettuces: nerves in shreds. We definitely deserved to celebrate. Not only had we got planning permission but also there was absolutely no mention of any untoward conditions attached to it. No reference to lime rendering, no reference to visibility splays, or any of the other onerous conditions that the Planning Officer had said she would be recommending. That woman had given us many sleepless nights. The only condition that had been applied was that the cider press be retained but, miracle of miracles, no mention was made of it being in its original position. Halle-bloody-lujah!

We could start in earnest now. There was nothing at all to hold us back. We started to work from dawn till dusk with an unrivalled enthusiasm. There was a huge amount of work to get through but we knew we could do it – no problem. We were invincible!

We had already approached Dawson, a local cabinetmaker, who we'd met in the Travellers Rest pub, about a hand-made kitchen. He was also willing to tackle the window- and door-frames for us. All the window and door openings in the barn were of an irregular size, and it would have been nigh on impossible to get mass-produced frames to fit. Henry, an extremely likeable chap, who had his own carpentry business but also often helped Dawson, came along with him to measure up. After talking to them both, we came to the conclusion that, although we would dearly have loved to fit the barn out with oak, the cost would be prohibitive, consequently, instructed them to proceed with the order in pine. We would then stain the frames and doors with an oak finish, hopefully achieving more or less the same look.

We had decided that the floor of our new abode would be on three different levels. This was not only to add character to the new structure, but also to mirror the original floor level, which followed the gradient of the lane outside. The level at the top end of the barn was a good three feet higher than at the lower end. We hired a mini digger to cut out a lot of the hard work and, with this intrepid little machine, Gordon spent more than a week getting the levels just right. As soon as the digging was completed, the concrete was ordered. We were to have a total of three, six cubic metre loads, and Gordon was to be aided by the two brickies, their labourer and - last but not least - me.

The first load arrived at eight o'clock on the dot. We were all ready and waiting, shovels in hands and a barrow for every man. We were fortunate in finding a concrete company who had a mixer truck with an extendable conveyor belt. The truck manoeuvred as closely as it could to the large opening, about half way along the middle section. The conveyor arm was then extended through the opening and managed to reach almost to the far side of that section. The topmost level of the barn floor was to be laid first but, as this was further than the conveyor belt could reach, all the concrete had to be barrowed up. The men acted almost like conveyor belts themselves, with one barrow always beneath the mixer truck arm, being filled to overflowing. It was a hard slog to push the overloaded barrow, across the scaffold boards that had been set on the dirt floor, and up to the far end of the barn. It was relentlessly backbreaking work.

When the first mixer truck was at last empty, there was still the raking and levelling to do. This was at least something I could help with and I did my fair share of the raking. There was just an hour between trucks, in which to finish off, and have a break. I was, of course, on tea duties as well as raking and shovelling.

While the men were sitting with a cup of tea after that first load, our neighbour, Brian, from the first of the two bungalows further up the lane, came to see what we were up to. By the end of the day, he must have wished he'd stayed at home! When he could see the enormous assignment unfolding, and the volume of work that was still to do, he offered his help. He was, obviously, welcomed with open arms. Even with five-and-a-half men, it was still a very long day. When the final truck had departed, we were all grey faced with fatigue, and covered from head to foot in concrete splashes. But the new floor was perfectly laid and one of the biggest jobs was now out of the way. We treated everyone to a well-earned restaurant meal that evening.

Unfortunately, we hadn't heard the last of the Planners. Early one morning we found a young man wandering around inside the barn. He was smartly dressed, in a suit and tie, so had obviously come for some official reason or other. This was probably why we were polite!

The young man said, 'Good morning, I'm the Bat Officer.'

We both gawked at one another, wondering whether we'd heard him correctly – Bat Officer? Come again? *Was this a wind up or what?*

'Good morning, what can I do for you,' replied Gordon, after doing a mental double take.

'I saw that you'd recently been granted planning permission on a cob barn, so came to see what provisions you've made for bats and owls.'

'Erm...well...as far as I'm aware, there are no bats or owls, so we haven't made any provisions for them,' replied Gordon puzzled.

'I can assure you that there would have been both here at one time. Bats are now a protected species and it's against the law, you know, to disrupt their habitat,' the young man said reprovingly.

'Yes, I was aware of that, but there are neither bats nor owls nesting here now. We haven't even come across any old nests to support the fact that there ever were any,' insisted Gordon.

'Nevertheless, provision must be made, just in case they should return, or others decide to take up residence here,' the pedantic young man said.

'Mmmm...what sort of thing are we talking about here?' asked Gordon, giving in to the inevitable.

'Well, for instance, some kind of box set into the roof is a good idea. We can come up with all sorts of designs for your architect to play with. I can let you have some plans when I'm next in the area, if you like.'

'A box you say? How big are we talking now...? Something like a wardrobe maybe?' enquired Gordon, not daring to glance in my direction, and deliberately ignoring my attempts not to laugh.

'Yes, a wardrobe would be an excellent habitat,' declared the young man, evidently delighted with our enthusiasm for his cause. 'Just make sure the access is large enough and they'll be very happy with that.'

'I think we've got one hanging around somewhere. I'll see what I can do.'

'Excellent...excellent. I'll drop round again, when I'm next in the area, to see how you're progressing.'

Shaking hands all round, the young man left, well pleased with his morning's work. We were left silently shaking with suppressed mirth.

CHAPTER SEVEN
COMING UP ROSES

The front garden of the original farm looked, like the rest of the house and yard, distinctly unkempt. There was a short crumbling stone wall surrounding the garden, topped by a hedge that appeared to be made up mainly of overgrown brambles and trailing honeysuckle. Inside the sheltering walls there was a straggly, patchy lawn, with a large circular cut out in the middle that was full of weeds and unidentifiable brown shrivelled things. In the borders around the lawn there were roses. Masses and masses of roses. Most were very well pruned, but you could see that they were old roses, knobbly and gnarled.

During the early spring, there had been little sign of activity in the garden, but the last few days of May had seen James busy with his spade edging the lawn, and down on his knees weeding. The empty containers of lawn fertiliser piled up and huge mounds of manure were forked in. The lawn was cut to within an inch of its now pristine life.

James was a sparse bantam of a man: not an ounce of fat on him. When he was all dressed up and ready to go into town, he looked more like a bookie than a farmer. All the Luscombe lads wore the same outfits when they went into town. The first time I saw James in his finery, I had to try desperately not to laugh. His sparse hair, with its bullet straight centre parting, was slicked darkly down with water, and in his hand was a jaunty little hat, with a green feather sticking out of the band. His suit was made of some sort of tweedy stuff - all thick and hairy - which might have been comforting in the winter, but he wore this outfit every single day - whatever the weather. I could foresee many sweaty summer days to come. His fawn waistcoat had five missing buttons, which had been replaced with string, tied into the tiniest and neatest of bows. And, fascinatingly of all, so were his flies! Obviously he was wearing the old fashioned type of trousers which would, once upon a time, have had buttons, but it seemed that these were also missing. Just peeping out, were three minuscule bows. How on earth did he manage to go to the toilet? I spent many amused hours imagining him attempting to wrest himself of this garment in the pub toilet. That must be the reason he drank whisky, rather than pints of best bitter!

He was working hard in the garden that particular day, wearing his usual working gear – a shirt so thin from constant wear that it was almost transparent - rolled up to the elbows, over the top of an extremely grubby looking vest. I had only seen grey combinations - the kind with a buttoned back flap - on the washing line, during their once monthly washing sessions, so I had to assume that instead of a vest, in all probability, it was an all-in-one. His trousers were dark with a greasy appearance, and might once have belonged to a suit. They had obviously long since been demoted; now showing tattered holes and patches on the knees.

As I was closing the gate on my way into town, James hailed me. 'Today's the day, maid!' He was struggling violently with his teeth, which were doing a clickerty-clicking action as they moved freely around his mouth.

I stood back a little, so that the tiny droplets of spittle that were spraying from him, wouldn't reach me. Also the gamey smell drifting my way was more than a little unpleasant. 'Yes?' I said, guardedly, 'what for?'

'Getting me plants in o'course,' he replied scathingly. 'It's the first day of June. That's when they all'is go in.'

He obviously thought me a total twit of a townie, that didn't know one end of a plant from the other. I had planted my own tubs, together with a few bedding plants, at least three weeks before; which obviously was not the correct thing to do at all.

71

'I lays it out just the same every year, so that sister can see it from 'er window'. As he spoke he gestured to the window behind him, to where I could just see a shadowy figure, peering through the grimy dust-rimed glass.

'You don't wants to do un too early in case that there frost gets to un.'

'Is that right?' I asked.

'That *is* right,' he reiterated. 'I can all'is remember going to see this 'oman. Now...' he scratched his head in bewilderment, 'she didn't know nort about plants. She was putting all them there big plants at the front, and all the little uns at the back!' He gasped and his eyes were rolling in his head. 'I kept tellin' 'er it were wrong but would she listen? She 'ad one of them there whatwheelers. She told us not to touch un but I'm that good with dogs. I knew'd it ud be all right. That there whatwheeler didn't bite me. No sir! He were a proper fool. Just came for a pat 'e did.'

Having listened to this tale on at least three occasions, I was well aware by now that a "whatwheeler" was a rottweiler. He did have an unfortunate tendency to repeat himself. Whenever either of us encountered James, and he was in the mood for a chat, the conversation would inevitably turn to his garden. He had told us, on more than one occasion, that people came for miles around to take photographs of the garden when it was in full bloom, during the summer. They would stand at the top of the road and snap away like some tourists visiting a Hollywood mansion. He had also told me, looking for all the world like a pouter pigeon, his chest expanding with immense pride, that someone had even done an oil painting of it, which was hanging over the mantelpiece in their sitting room.

The story of the lady and her rottweiler having run its course, he continued, 'Just you wait a minute, I'll just go and fetch that there picture, so's you can see what me garden looks like.' And off he went leaving me thinking of all the chores I still had to do, and how late I would be back home. Oh well...

He returned with a hefty ornate gold-framed picture of the farmhouse and garden. It must have measured at least four feet by three feet. On closer inspection it was obvious, to me at least, that it was a blown up photograph which had been printed on textured paper but, the Grimms, simple souls that they were, thought it was a real oil painting that had taken weeks to do. It had also, more than probably, cost them an arm and a leg. Or maybe, come to think of it, a more likely scenario was that they had spoofed for it. This was a game they loved to play whenever they got the chance. They would grab a handful of coins from their pocket and ask you to guess how many they had in their hand. I think the object of the exercise was, that if you didn't guess correctly, you had to give them the equivalent amount of coins, but they would play the game with stones, or anything else to hand, if they were short on coins.

Both James and George both liked to "get one over" on us whenever they possibly could, and did their level best to get something for nothing. Often we'd be asked to run errands for them, and there always seemed to be plenty of those, despite the fact that they went out most days themselves. Gas was one particular favourite. Bottled gas, that is. They had a mobile gas heater but they obviously didn't know how to disconnect the bottle from the appliance because, whenever it ran out of gas, they would drag the whole shooting match through the front door, all the way through the garden, to the little gate at the side out into the lane. Here it would sit until Gordon spied it, disconnect the bottle from the body of the fire, and then heave the large heavy cylinder into the back of the Land Rover.

Mostly the shopping fell to me, so the replacement gas cylinder was also down to me. If I hadn't intended going shopping that day, then it was just too bad as I'd get a real guilt trip thinking of them sat huddled together with no heating, during those still chilly evenings. This mobile gas heater was the only form of heating in the farmhouse as, amazingly, although they had chimneys, they didn't seem to use them. The only place I knew where to get the gas bottle replaced was at the very far side of the industrial estate, on the outskirts of Exeter, way off my normal route. I would be chuntering and muttering all the way there and back about the Grimms.

On my return from town, the new bottle would be reconnected to the fire, by Gordon, and eventually dragged back the way it had come, by one of the Grimms. Only later on in the week would there be mention of payment.

''Ow much do we owe ee for that there gas maid?' George would say.

'Oh, it was eight pounds sixty pence,' I'd pipe up, knowing full well they would wriggle like hell to get out of paying.

'What say we call it double or quits then?' he'd say hopefully.

'No way, George! You know I wouldn't gamble with you – you always win - pay up like a gentleman.'

He would only then, sighing mightily and shaking his head in resignation at my lack of gamesmanship, put his hand in his grubby pocket and pull out a pile of unpleasantly warm coins.

Standing in the lane with the enlarged photograph in my arms, I said, 'Nice, James, very nice. You must be very proud of this painting, and I can't wait to see the garden when it's finished.' I handed him back the picture.

'I does it very nice, don't you think? Allis the same every single year.'

'Yes, it must feel very satisfying, creating all that beauty out of nothing. Well, I must dash now, though, see you later!'

When I returned in the late afternoon, I could see he had been hard at it. The empty circular bed now had regimented rows of plants, and it looked like each one had been laid out with a ruler. The outer edge was planted with clumps of white lobelia, next came a ring of purple pansies with yellow centres, followed by an ordered circle of golden marigolds. The very centre was filled with dahlias in a riot of mixed colours. An instant garden. The roses that were all round the edge of the garden had been flowering unnoticed for some weeks now and the whole effect was totally enchanting.

Unfortunately, Billy Goat Gruff and his family reciprocated my thoughts. They must have been able to smell the tasty new plants and somehow made their way from the field, presumably via the streambed, and over the wall into the garden. When James saw them, all hell broke loose. He rushed out of the front door as if the hounds of hell were after him, screaming and shouting and waving his hat in the air. The goats were not to be shifted that easily, however, and it took him a while, with the aid of George's stout stick to persuade them to leave. They didn't go too far down the lane, though, and it looked to me very much as if they would resume exactly where they left off, just as soon as they possibly could. James obviously saw the same look in their eye. When he saw me watching, he came over, 'I needs to get rid of them bleddy beasts. Just look what they've done to me garden!' He was very agitated, spit and splother raining everywhere.

'They've made a bit of a mess but I think you've caught them in time. It's mostly just a few leaves James, I don't think the plants themselves look damaged,' I placated.

'I'm going to shoot the buggers, see if I don't!' he said, not easily appeased.

'Can't you stake the goats for now, to stop them getting back in?' I knew if he could just calm down things would be OK for the goats, he was very soft with animals normally, but his beloved garden had suffered a severe blow. He had collected a broom from behind the tree and started to sweep up some of the mess. I think he could see that disaster had been averted and he started to visibly relax.

'I've had an idea.' I said, suddenly having a brilliant thought. At the Travellers Rest they were just having a beer garden constructed, with a play area for children. 'Why don't I ask Paul at the Travellers Rest if he wants some goats for his new beer garden? He could put them in a little paddock next to it.'

'Why that's a proper job!' His face brightened at the thought and he slapped his hat a few times against his thigh. 'Them kiddies ul love 'em. What a good idea!'

He went off to stake the goats, while I went in search of my mate to tell him the glad tidings that we had a mercy mission to run that evening. We certainly had a very enjoyable evening and, after a chat, Paul, the landlord, was all in favour of taking the goats. A few days later they were transported with the aid of a borrowed horsebox down to their new quarters, where they were greatly admired by the children all summer.

The progress on the barn had been rapid since the floor slab had gone down. The two bricklayers and their labourer were on site most days, beavering away, building the internal block-work for the new roof to sit on. There had been a bit of an argie bargie going on between the Listed Building Case Inspector and the Building Regulation Officer. Letters and memos had been passed and each of them had visited the site and commented. The Listed Building man wanted the internal walls to be left as they were – just cob – with a lime-washed finish. The Building Regulations Officer didn't agree, he wanted an internal skin of block, leaving a cavity between. We were on the side of the Building Regulations Officer. We wanted to sit the new roof structure on the top of the new internal walls for strength, as the cob walls alone wouldn't be man enough for the job. Having a cavity would also give us the benefit of insulation and damp proofing. In any case, a great deal of the cob had previously been repaired in block and brick over the centuries: the lower section, where the milking parlour used to be, was already rendered with sand and cement. This made the arguments put forward by the Listed Building Case Officer, unenforceable. In any event, he had to retire from the case when he discovered, to his chagrin, that although the Grimms' house was Grade II listed, and the same age as our barn, the cob barn roof was not the original sixteenth century structure. This meant that he was unable to list the barn or have any further say in our case. Goodie! We were left, breathing a heartfelt sigh of relief, to get on with our original plan of having a cavity, thus in effect giving us a new house within the skin of an ancient building.

We had decided to have only one storey at the top end of the barn, which was destined to be the sitting room. The ceiling would follow the curve of the roof, leaving the four massive oak beams that tied the two long walls together, floating in mid-air. Most of the beams were in reasonable condition, with minimal damage from woodworm. The beams were oak and, in places, they even had their original bark still on them. It was my task to clean them up, before a professional company came to spray them with preservative. Gordon set a small scaffold up for me to climb on and off. The beams were probably about two feet thick and the easiest way to clean them up, I found, was to straddle them. Very nerve racking for me, as I absolutely detest heights, but apart from that the job was well within my capabilities, so my fears had to be ignored. I found, by just concentrating on the stretch of timber in front of me, avoiding the gaping chasm on either side, that I could work well enough.

I first chipped all the bark off with a hammer and chisel, and then smoothed any rough patches with coarse sandpaper. There were quite a few ancient hand-made nails embedded in the beams, some of which defied removal, so were left in place to give a bit of character. After my ministrations, the beams looked very acceptable and were duly sprayed by a preservation company against woodworm. After they had been left to dry out for a few days, I then varnished them to a soft satin finish. I had wanted to treat them with bees wax, but thought the varnish option was the more practical solution. They were a long way off the ground and, even with a ladder, it would be difficult to reach them once the scaffolding was removed. Later on, we were given a recipe by the Conservation Officer for sealing historic timbers, unfortunately not in time for me to apply it to our beams. This consisted of two parts turpentine to one part raw linseed oil; to every litre of which you then added two hundred millilitres of melted bees wax.

Meanwhile, James's garden continued to thrive. It was laboriously watered every evening with a small metal watering can, even if there had been rain that day. The lawn was immaculately cut every Sunday morning - rain or shine. Every now and again there would be a hiccup with the lawn mower. It was a rather ancient electric mower, which always seemed to pack up half way round the lawn leaving James bending over it, muttering to himself, a probing finger lifting the edge of his hat to scratch his head. As none of the other brothers seemed any more mechanical than he was, he would then have to come to see if Gordon would lend him a hand to fix it. Having to ask Gordon a favour of any kind was not something he relished; something to be avoided at all costs.

The marathon drinking sessions they indulged in during the week would, often as not, lead to a row between Gordon and either James or George, or, both James *and* George. This would sometimes be followed by a period when they would totally ignore our existence. I think the severity of the cold-shoulder treatment had some bearing on exactly how much of the row they actually remembered next day: whether they recalled Gordon's insults, and whether they struck home. Anyway, if the breakdown in the mower coincided with a "not speaking" period, then James would find himself in a cleft stick. But the Sunday ritual, and his beloved lawn, would eventually force him to swallow his pride and come cap in hand for help.

For a man born and brought up on a farm, his lack of mechanical adeptness seemed puzzling. As far as I could see, none of the brothers appeared to be any better. When we first met them, they had no vehicles of any description - not even a tractor - which was surely very unusual on a farm: especially when you bear in mind that they originally had in excess of a hundred acres to work. Apparently not long before we arrived and, after a particularly heavy lunchtime session, they had crashed and rolled their car over into a ditch. They were very very lucky that none of them was hurt although the car was a total write-off. James was also fortunate not to have his driving licence taken away but presumably, as they were all elderly and the accident happened during daylight hours, no one realised what a drunken state they were in. Neighbours told us that Social Services provided the car in the first place, as Verity was disabled, and it took quite a few months for them to organise a replacement for the crashed vehicle.

James had been driving the new car for at least a month when I saw it in the lane, with the emergency lights flashing. I had seen him walking round the vehicle a couple of times, as I approached down the lane, returning from a walk with the dogs. He was scratching his head and looking a bit puzzled, so I asked him what was the matter.

'Buggered if I know maid,' he replied, tipping his hat.

'Well, I'm not sure I'm going to be of any help to you: cars are a complete mystery to me. I don't even know where the bonnet catch is on ours.'

'Tis a proper mystery and no mistake.'

'What's wrong with it?' I asked, hoping for a little more information.

'Dunno, I'm sure. This 'ere light just keeps flashing and I can't stop it. Look 'ere...' He caught my hand and took me to see the two flashing lights on the front of the car, swiftly round the back to see the other two, and then pointed inside the open window to the console above the steering wheel, where the red emergency button was flashing on and off.

George sat grinning in the back with two grubby looking West Highland terriers on his knee, while Verity was in the front, her legs covered in a tartan rug. I leaned inside, pressing the emergency button once whereupon the flashing stopped.

'You've fixed it!' he gaped at me in amazement and some embarrassment. Amazement because I was a mere woman who could do miracles with a car: embarrassment, because he'd had to ask me a favour. 'We'd best be on us way then, us is late for market already,' he gabbled. And with that parting shot, he climbed back behind the wheel and juddered up the lane.

The mower continued to struggle on but there came a day when even Gordon couldn't get the mower started, and a

Sunday had to go by without the lawn being mowed. Obviously the time had come to replace it, and a few days later we saw James struggling to unload an enormous green and red mowing machine, which looked as if it would take two men to push. He manoeuvred this hulking great motor mower onto the lawn, and this is where it stayed for another week, while he fiddled and pushed it around. It became obvious he couldn't even start it!

We had been avoiding the Grimms for days now, watching, like two naughty children, with unholy glee as the lawn got longer and longer. We had seen James trying unsuccessfully to start the new mower; getting hotter and hotter under the collar, in his efforts. His temper was never very far from the surface at the best of times, but after his usual session in the pub after doing the shopping, and his late afternoons spent trying to start the mower, this did not, surprisingly, improve. You could almost see the pressure building in him, day by day. The problem was, that he had to steel himself to ask for help again. With the whisky freely flowing through his veins, he just couldn't bring himself to ask a "mushroom"!

'Nice mower, James, should cut the lawn a treat,' Gordon would taunt, grinning from ear to ear, in anticipation of another row to come.

It went quiet for a few days and then, to our immense surprise, we heard a loud engine noise that could only be coming from the new mower. Poking our nosy heads round the barn wall, we saw Mrs Nick Nock whizzing round the lawn at the furious pace.

It was a long time before we found out that Mrs Nick Nock's name was actually Mrs Stick-Lott but, by that time, the nickname had firmly stuck. The Luscombes were either unable to pronounce this unusual double-barrelled name, or more likely, had made up a fatuous name that appealed to their strange sense of humour. Mrs Nick Nock was married to a local businessman. They were both of them well into horse riding and hunting, and lived in a house further down the valley. I was not sure of her connection to the Luscombes but supposed that, as a near neighbour, they were friendly. They were the only people we had so far seen invited inside the house, which marked them immediately as extraordinary.

Mrs Nick Nock usually came alone to visit the Luscombes: she came most often early on Friday evenings, parking her Range Rover at the top of the lane. She would spend quite a few hours chatting over the wall, usually to George, who always gave me the impression he was flirting outrageously with her. He thought himself a proper old dog. Whenever he'd been having a chat to her, he was always in the best of moods.

Mrs Nick Nock was only a very slight woman, not much more than five feet tall, dressed smartly in her usual riding habit, hanging onto the handle of the colossal motor mower for all she was worth. It looked very much like the mower had a life of its own and was certainly in charge of their canter round the garden.

The Sundays that followed found a succession of different people manoeuvringing the mower round the lawn, followed by James plying them with liberal measures of whisky. Even Gordon took his turn, when a benign mood was on him.

We, meanwhile, had been getting everything as ready as possible to start the new roof structure. Gordon had been doing as much preparation as was practicable, before the old roof was stripped off. He had cut a total of one- hundred-and-eighty rafters and these were now waiting inside the Atcost barn. He was going to start constructing the new roof under cover of the existing one, as the height difference, from one end of the barn to the other, was a massive three feet. It was extremely important not to get any of the cob wet, so we were intending to split it into sections and do each, wherever possible, under cover of the existing roof. Each section would have its new roof timbers pitched, felted and battened, before the old roof would be removed from above it.

We had decided that we were going to try and re-use the existing terracotta roof tiles with their lovely aged patina, which would take new roof tiles many years to replicate. We were going to have to take them off very gently, to save as many of them as possible, but had luckily sourced a reclamation yard that we could purchase some spares from, if necessary. We were praying for a spell of settled weather. We would need at least three weeks without rain for the job to be completed without mishap. Probably asking a little too much of the British weather.

The summer was progressing nicely when our neighbour, Steve (who lived in another conversion of a barn that had originally belonged to the Grimms), departed for his usual two month holiday at his second home in Cornwall. He was the one most often on the tiller of the new motor mower, and would be sorely missed by the Grimms. Steve had always got on reasonably well with the Luscombes. He had purchased his barn from them when they had first started to run out of money, replenishing their resources, at least temporarily - unlike us - who had only filled the coffers of the Receiver.

Quite a few Sundays went by without a sign of the grass being cut, although James could be seen doing his usual stint of weeding and deadheading. Where were all their usual slaves? 'What's up?' we were asking ourselves.

That Sunday we were up early: six o'clock. The weather had been very warm for the last few days and we thought we'd get an early start and then pack up at lunchtime to enjoy a well-earned afternoon off. It was Sunday after all. Following a hurried breakfast, we trooped over to make a start on cutting yet more rafters. On our way from the "big" barn where the caravan was parked, we saw James down on his knees on the dew soaked lawn, cutting away with a pair of ancient sheers. We both came to a standstill absolutely riveted by what we saw. He was clipping his way across the lawn. The sheers making a sharp snip, snip, snip sound. After he had laboriously completed one small section to his satisfaction, he then picked up a pair of scissors and cut precisely along the edge! Our mouths hung open as we watched in sheer amazement. James's dedication to his garden knew no bounds.

That evening I stood in the yard looking up into the ink-black sky. It was a vast dome over my head filled with twinkling stars. The enormity of it engulfed me, and I stared at it awe struck. I was able to pick out the plough, but was a complete ignoramus as to the rest of them. There wasn't a sound of human activity, only a lone owl hooting. What a precious thing silence is. Gradually I realised that the warm still air was full of the scent of the Grimms' roses. Dozens of different perfumes, intermingling to delight my senses. Taking the time to appreciate my surroundings hadn't been top priority up to now, but I made a pact with myself there and then to make more time in future.

CHAPTER EIGHT
RELIGIOUS ENCOUNTERS

July started hot and dry. The weather forecast had been consulted feverishly, like the oracle, for weeks now. We needed a good settled period of dry weather and it looked as if we were in for a spell of just that. The decision was made; we would start work on the roof next day.

We had made enquiries locally for a labourer to help us while we did this part of the job. Not only was there the problem of the roof having to be completed in the least possible time to avoid the cob walls getting wet, but also as I had a phobia about heights, Gordon wasn't quite sure how much help I was going to be. He had already encouraged me to climb the scaffolding a few times, trying to get me used to being on the top of this precarious structure. He'd climb right behind me, reassuring me every step of the way. Although I will admit it was a large platform when you actually got to the top, it was extremely nerve racking getting that far. It was necessary to climb from the first ladder onto only one small scaffold board and then shuffle around onto yet another ladder, before reaching that top platform. I would shiver and shake for half-an-hour after I had made the mammoth attempt at climbing, what appeared to me, more akin to the ascent of Mount Everest, sad specimen that I was.

We didn't want to commit ourselves to employing someone full-time because, to be honest, we just couldn't afford it. We were looking for someone who was either between jobs, or just wanted a working holiday. Not surprisingly, this person proved very elusive. Taking pity on us, the landlord of the Travellers Rest pub, who also had his own building company, loaned us a labourer for four weeks.

Peter duly arrived for work next day: promptly at eight o'clock. A good start, but problems were encountered almost immediately. He was scared of heights too! Knowing his weakness, Paul, his boss, hadn't told him that he would be helping to construct a new roof. What a rotter! Poor boy went as white as a sheet when he was given his orders for the day. Gordon was not impressed when he had to coax the two of us up the scaffolding, and treated us more like a couple of kids he'd been left in charge of, than co-workers. After the tremors and shakes had left our limbs, Peter and I started stripping off the old tiles, while Gordon got on with assembling his rafters, on the piece of roof he was constructing, under cover of the old one.

Peter and I worked well as a team and "the boss" was eventually pleased with our progress. We soon got into a routine of stripping one section at a time, with Gordon following on behind with his rafters. We would all then lend a hand, felting and battening, before moving on to strip the next section.

At the end of the week we had completed about half the roof and the weather still looked settled, thank God. I had offered up many a prayer for a dry spell, and it perhaps wasn't too difficult to understand why our conversations, in our lunch breaks, seemed to drift on to religion. Perhaps working on the roof, and the elements being so vital in the scheme of things, had had this effect. Peter had some odd ideas about religion in general, but our discussions had reminded us of an incident that had happened to us here in the yard a few months before, and on another occasion just a few years ago.

Our first encounter of the religious kind happened one rare hot sunny day in May. It had been raining cats and dogs for days but we had managed to find jobs indoors to keep out of the wet and cold. On this particular day, though, the weather had miraculously cleared and we were outside having a well-earned cup of tea, just glorying in the warmth of the sun. We were sat on an old teak bench that we had set up against the wall of the barn, in a nice sheltered spot, that was not overlooked by the main road or our own lane. Totally and blissfully private. Even Ted wasn't around to plague us, as we'd seen all of them leave in the car. This was a rare day when we had the site to ourselves and one we were both determined to make the most of.

The wall behind us intensified the strength of the rare sunshine: it got so hot that Gordon had stripped off entirely, determined on an all over tan this year. He sat there very contentedly, with his face turned up to the sun, slowly sipping and savouring his cup of tea. Not to be outdone, I decided on the all-over approach myself and had soon joined him, raising my arms lazily above my head to let the rays of the sun feast on my body.

A few minutes later the dogs started a frenzy of barking, alerting us to the fact that there were strangers around. They didn't make that kind of fuss when it was one of the Grimms, as they knew them all by now, but we had in any case seen the Grimms leave a few minutes before we started tea break. We had also seen both sets of neighbours from the bungalows up the lane go out and had thought we had the place to ourselves. As our little complex of dwellings was at least two miles from the next habitation, it was usual for anyone visiting to arrive by car. We hadn't heard the sound of a vehicle approaching; just ignored the dogs, thinking it must be a fox that had spooked them, and so continued to revel in our unexpected holiday.

Gordon had started to get that certain twinkle in his eye that had been stimulated by the sight of my sun-kissed breasts, when we were dumbfounded by the apparition of two middle-aged ladies calmly entering through our five-barred gate and into the yard. Strangers. Both were smartly dressed in flowered skirts with crisp white blouses; one lady was extremely fat and the other painfully thin, both clutching armfuls of books and pamphlets. Neither Gordon nor I had dared move a muscle and we watched in utter fascination as they approached. We were like two rabbits caught in the glaring headlights of a car.

'Can I help you?' said Gordon as cool as a cucumber. Both ladies stood transfixed, mouths agape taking in the scene of debauchery before them.

'Not dressed in that fashion young man!' The large one boomed witheringly. She threw her newspaper, accurately covering Gordon's interesting bits, before grabbing hold of her colleague's arm and swiftly dragging her back in the direction of the gate. Both of them leaving us in no doubt that they were totally appalled at the sight they had witnessed. She clanged the gate shut with an angry bang and marched emphatically back up the drive, with her associate scurrying to catch up. They must have walked along the main road, for there was no car waiting for them at the top of the lane. They were in for a hot, dusty walk to the next farm. I could only hope that their reception at their next port of call was not so shocking as that endured in our yard.

The telling of that little episode gave us all a good laugh and then reminded me of another incident that had happened a few years before. We were between barn conversions and had been renovating a small house in a market town, not too far from where we were then living. The day the incident happened, I had been stripping the wallpaper from the small front sitting room and had then started sanding the woodwork in the bay window, ready for painting the next day.

The window overlooked a dark and gloomy little overgrown garden. I had been daydreaming about clearing all the overgrown weeds and rubbish out; imagining what sort of shrubs I would start putting in, once the job had progressed that far. Two earnest looking young men, letting themselves in the front gate, interrupted my reverie. They were dressed alike in navy blue suits, crisp white shirts and matching ties. A couple of clones. They knocked importantly on the front door.

I was just about to go and find out what they wanted when, on walking round the door into the hall, I saw that Gordon was already there. He had been painting the back of the front door and was at that moment on his knees, finishing off the bottom section under the letterbox. Instead of getting up to opening the door, he lifted up the flap of the letterbox and said, 'Can I help you?'

He seems to use this little phrase quite often, getting some weird and wonderful answers, in some equally weird and wonderful situations. This time, I didn't actually hear the answer but could see the result, as I had, by this time, returned to my sanding job at the window. I could see one of the young men kneeling down on the path, oblivious of the damage he must have been doing to his trousers, looking through the other-side of the letterbox.

To this day, I cannot imagine what on earth he thought the man on the other side of the door was doing on his knees, talking to him through the letterbox. Did he perhaps think he had stumbled on someone praying for divine intervention? Was this something the locals did on Saturday afternoon to pass the time? Gordon told me later that he had said, 'I'll pray for you,' as he left. I just hope that the poor man couldn't hear the hysterical laughter coming from the bay window, as he went on his way. The tears rolled down my cheeks, as I literally fell on the floor, doubled up with laughter. Gordon was in no better state. It took many hours before I could summon the energy to carry on with the job, after that debilitating mirth.

If any of the people mentioned in the above episodes should ever read these tales and recognises themselves - truly, I am sorry!

The work on the roof, miraculously, was only interrupted by rain on one occasion. Fortunately, we were quick enough to get a tarpaulin over the cob, in time to stop it getting a soaking. We were very lucky that the rain mainly held itself in check, until the whole roof was felted and battened. It was certainly asking too much to get the tiles on before it started! The task had been an arduous one, leaving us all feeling exhausted and ready for a long holiday. Not only was the roof more than one-hundred feet long, but we also had a total of eight Velux roof windows to incorporate. Very time consuming, fiddly work.

The tiling we were able to do at a more leisurely pace. The pressure was off. We were fortunate that the majority of the tiles had been removed in good condition, only needing one or two replacements. I was even careful when I gave them a brush down, by not removing all of the old ochre coloured lichen. When the roof was finally finished, it looked as if it had been there for centuries - just what we'd hoped.

The roof at the end the four weeks was still not totally completed. There were still fascias, bargeboards, soffits, and guttering to finish off, but these could safely be left until other more important tasks had been caught up on. Peter, our loaned labourer, was sent on his way with a full wage packet and grateful thanks from both of us. We would have struggled without him and no mistake.

When the new roof was done and dusted, it was time to start a job we had been avoiding. The barn would not be on mains drainage, which meant we would have to construct our own septic tank. With a stream dissecting our land, and much of the surrounding area already concreted over, it became obvious that the only place to install it was in the old farm slurry pit that was sited between the bridge and our main gate. This was an enormous open-topped structure: built of reinforced concrete. The whole of it was surrounded by barbed wire to stop people, and animals falling in, as it was probably more than ten feet deep.

The reason we had been putting off the evil day, was that it was still full of black sludge. This had been the main slurry pit for the farm and had taken the run-off of the liquefied urine and dung from the farmyard. It was illegal to allow this slurry to enter a watercourse, as it was a pollutant and deadly to fish, so we knew the tank must be watertight. If this had been a normal farm, then probably this slurry would already have been pumped into a sludge gulper, then sprayed on the fields to fertilise them. As it was, this evil looking sludge was still lurking in the pit.

Gordon set off late one afternoon to walk along the right of way to Rick's farm, to ask if he would pump out the slurry for us. Rick was a great conversationalist: always willing to indulge in a good chinwag. Gordon was inevitably gone for some time, but Rick proved more than willing to help us, arriving next day with his tractor and gulper.

It took him three trips with the tanker to extract the foul concoction in the pit. He took it up to the top of the orchard and gave our grass a liberal coating of the evil mess. After he had left, there was black gunge just about everywhere. It was all over the yard and there was a trail of it up past the Atcost building where the chicken house was. The dogs and, for that matter, the chickens were going to be covered in this stuff for weeks, unless it rained pretty soon.

The only problem now, was that Rick's pump couldn't suck out all of the sludge in the pit, which meant that next day, a reluctant Gordon, kitted out in a boiler suit, gloves and long wellies, climbed down a ladder into the glutinous black mess left in the bottom. He filled his bucket, climbed back up the ladder and emptied his repellent load into the front of a borrowed dumper truck. When the dumper bucket was full, he then drove it up on to the grass and emptied it. He did this for two whole days! What a hero.

When the pit was finally empty, he was then able to start work constructing the septic tank. He divided off the original tank and in one third of it built a three-stage traditional tank of blockwork and render. When the house would eventually be occupied, the liquid effluent created, would be pumped up from the holding tank to a large soakaway, in the far corner of the orchard. The other two-thirds of the slurry tank were then infilled with rubble, subsoil and finally topsoil. This fine place was where I would be sowing the seeds for our vegetable garden.

We had been so busy on our roof and septic tank, working as many hours as our bodies would take, that we hadn't noticed that Ted was not doing his usual rounds. When we were able to take stock, again and start noticing our surroundings, we missed his little chats, speculating on his absence. It was most unusual not to see him every few days.

I spotted James mooching around in the garden early one morning and asked him, 'Mornin' James, how's things? Haven't seen old Ted around lately, what's up?'

''E aint been well lately, that's what's up, maid. I'm waiting for doctor to come right now,' he replied.

'Oh...I am sorry to hear that. What's the matter with him?' I enquired.

''Is leg is all black and swollen like. It don't smell too good, neither, come to that,' he recounted dramatically.

Oh no, it sounded like gangrene. I could remember seeing Ted some weeks ago limping and, when I asked him what the matter was, he'd told me he'd knocked his knee while he'd been mucking out. He said it had been troubling him for some time and was pretty painful. I'd told him to take care of it and to get it cleaned up, without any real expectation that he'd do it. Their living arrangements appeared to be so basic that it was no wonder something like this had happened.

'That sounds really awful. Is there anything I can do?'

'Nothin' any o' us can do, 'cept per'aps doctor,' he replied.

'Well, please give Ted my regards and good wishes for a speedy recovery,' was all I could add. I felt quite depressed after that little exchange. They really were incapable of looking after themselves.

I didn't notice the doctor's car during the morning, but later that day, did see the ambulance that made it's way down the lane. Two burly men duly carried a stretcher in, manoeuvred Ted's bulk through the front door and out into the waiting ambulance. *It must be serious then.*

I didn't see any of the Grimms for a couple of days, despite looking out for them. When I spied George in the lane I hotfooted it over to him.

'How's Ted doing? Any better?'

"E's not too bright, poor owd chap,' George replied, aiming the usual gob of spit to the side of him.

'Poor old thing! Is there anything we can do? Do you think he'd like some fruit, or can you suggest something else I can get for him?'

'Don't trouble about anything missus, he's being well looked after by them there nurses.' He seemed unperturbed about his brother's health, proceeding to tell me just how nice he thought the nurses were. Dirty old man.

It was only a few days later that poor Ted died. James told Gordon that they'd been to visit Ted in the afternoon. He'd been bright enough and had chatted to them for an hour or so, but they'd only just arrived home, when they had a telephone call from the hospital to say that Ted had suddenly died. We never did find out exactly how he had died, as neither of the brothers seemed to know.

The funeral took place at the local village church and, although, we were still very busy, decided that we must both make an appearance. The beautiful old church was surprisingly packed. Although Ted didn't get many social callers to the farm, he was obviously very well thought of in the community. Poor old Ted. I would miss our chats.

CHAPTER NINE
THE MOTLEY CREW

As the summer gradually faded into autumn, so James's gardening problems disappeared. There was now very little need to cut the grass, as the weather had turned cooler, and it no longer had the impetus to grow with any fervour. We had not encountered that delightful sight of him attempting to cut the grass with scissors for some time.

There was now a distinct sharp nip in the air, first thing in the morning, and last thing at night. With the weather cooling off, our life in the caravan had begun to pall: I had started to count the days until we could move into the barn. We had set ourselves a target of a November moving date but, every now and then, began to doubt how realistic this date actually was.

My sister (who lived in the United States of America with her husband and two young children) had asked if she could come to spend Christmas with us, as we hadn't all been together for at least three years. This would also certainly mean that my parents (who lived on Alderney in the Channel Islands) would want to come too. So all five bedrooms would need to be completely finished and also furnished.

After a lot of soul searching, we had both agreed we could do it. Although the inside wouldn't by any means be finished for the festive season, we thought we could make the barn comfortable enough. On this understanding, our visitors were invited.

Progress had been reasonably swift so far, but the sheer volume of work still to be completed, was daunting. With a house full of guests due for Christmas, this was now putting us under more than a little pressure. We had an almost permanent workforce on site and it was a very rare day now for Gordon and I to be working on our own. For the whole of the summer we'd had two brickies and their labourer working most weekdays. Gordon, as carpenter, was doing all the woodwork himself, with me assisting in any way I could.

The first fix had already been completed with the electrical work, but the plumber had so far proved elusive. We had used the same plumber since we'd first started renovations and conversions and had got into the habit of give him a call at the very start of each job. We had nicknamed him Leaky some years ago as, although his plumbing work was faultlessly neat, it wasn't always watertight. We'd had many spectacular events, when streams of water had been gushing out of the front door, but we'd used his services so many times in the past, that by now he'd become a friend as well as a valuable part of our regular workforce. The thought of getting in quotes for the job, rather than just letting him get on with it, had never crossed our minds. He had already been out to visit the site, when we'd first taken it on, and we'd given him an expected start date for the first fix, as the end of July. The arrangements were made and cast iron, or so we'd thought. After many phone calls where we only got through to his answering machine, leaving endless messages, or he called us promising faithfully he would be with us soon, he at last dropped his bombshell. He had taken on too much work and wouldn't be starting our job at all. With the clues we had already garnered, we should have expected it, I suppose, but it just hadn't occurred to us that he would let us down. Especially at this late stage. We were frantic. The whole job was now being held up for lack of a plumber. Not only that, Christmas might have to be cancelled.

The yellow pages were hurriedly consulted but it appeared that every plumber we tried was knee deep in work. We were having a Rayburn cooker installed and, on looking through the literature that had arrived with our order confirmation, I saw that there was a list of authorised plumbers attached to it. There were actually two plumbers for our area, so Gordon gave both of them a ring to see if they could help us. We were not optimistic, as most good plumbers normally have a full order book, but the Gods were with us on this occasion and we managed to find one that had, that very day, had a large job postponed. He could help us. Our relief was palpable. Christmas was back on again.

All the internal blockwork was now in place, and Gordon was busy putting up the studwork for the internal partitions, for the rooms on the first floor. A gang of plasterers was working behind him, fixing plasterboard to his timber partitions, and then plastering walls, ceilings and blockwork, as soon as they became available. I was still working on the external scaffolding, staining the fascia boards a rich dark oak colour. Keeping busy and out of the way.

While all this activity was going on, the Grimms' dogs had been growing in number, unnoticed by us. They were like weeds in the garden. One day you would see just a few and the next you were knee deep in them. It was very difficult to say exactly how many there were because they all looked so similar. They were painfully thin dogs, with deep barrel chests striped by prominent ribs, and long gangly legs. Their ears stood erect and they had long whippy tails. They would slink around, stalking us evilly with their mean brown eyes and sharp snouts. I had never in my life seen such ugly dogs before. When James told me that they had once bred racing dogs it became obvious where the "breed" came from. Mix in a little bit of collie and you had the perfect Grimm dog.

Standing in the lane, leaning against the barn wall, James produced an old black and white photograph from his back pocket. 'See that? That there's blackie.' It was a whippet standing smugly by the side of a table, on top of which was a silver trophy cup.

'Yes, lovely dog, James,' I said, handing the photo back, trying frantically to carry on with my job of staining the door and frame of the front porch. Out in the lane I was a sitting target for his reminiscences, and could do no more than half listen politely, while I carried on work.

''E were a good un. Won lots of races 'e did. Won lots of money on 'im too! Wish I 'ad a dog now 'alf as good as 'e was,' he rambled on, oblivious to my disinterest.

'You've got quite a few dogs now, though, haven't you?' I said after a short silence had fallen.

'Yes, that *is* right.'

'Just how many have you got, anyway?' I asked, curiously.

'Dunno I'm sure. Maybe six, maybe seven, can't rightly say,' he replied, just as if the question had never occurred to him before.

I knew that his estimate was nowhere near the mark. It was quite easy, even from where I was standing, to count more than six or seven dogs in my line of sight. And.... let's not forget the two Westies that were kept in the house. I also had an inkling that there might be yet more dogs, being kept in the deep litter house, which would once upon a time have been used to house poultry, on the far side of their top field. I could often hear dogs barking from that direction, and it certainly sounded like more than a couple.

'Don't you know from their names how many there are? What are they called?'

'Well, that there one is blackie.' He pointed at a muddy-black coloured dog that was standing by the gate. 'And this one 'ere is brownie.' The dog in question was fairly similar, a sludge-coloured brown dog, sniffing at his ankles. 'We calls all of 'em either blackie or brownie. It's easier that way,' he said logically.

I could see his point, really: they all looked as if they came from the same family. They had very few distinguishing marks: even the colour of them was hard to recognise, being neither totally brown nor totally black. Actually, now I come to think of it, a hard decision whether they were a brownie or a blackie, I would have said.

I could see another couple of dogs, mooching around by the top gate with, what appeared to be, a tribe of tiny puppies. 'It looks like you've got some pups up by the gate there,' I said.

'Ahh, so we 'ave. Nice little fellers, aint they? Wouldn't you like to 'ave a couple of 'em? They'd be good guard dogs they would,' he boasted, the salesman in him coming to the fore.

'Yes, they do look lovely but we've already got two dogs, and it's getting a bit cramped in the caravan as it is. I'm not sure we've really got room for any more.'

'Oh, you wouldn't keep 'em inside, maid. You'd have to make a little old kennel for 'em in the yard and they'd be snug as bugs.'

I was still trying to come up with a convincing argument against being lumbered with a blackie or a brownie pup, when he walked off into his own yard, appearing a couple of minutes later with three puppies squirming in his arms. Obviously there were more puppies than I thought!

'Now, 'ow could you resist these little 'uns?' James said, at the same time dumping one of the pups into my arms. He was fat and fluffy and so far, at least, showing no signs of how he would look later in adult life.

'Ahhh, isn't he sweet,' I cooed. That was just before I spotted something crawling on its snout. What was that...? *Oh, God, it's got fleas*. I handed it hastily back. 'What a little cutey, have him back quick before I get too attached to it.' I diplomatically said.

'You could easy keep a couple in that there yard. They'd be no trouble,' he cajoled.

'Have you found homes for any of them yet?' I asked hopefully.

'No, not yet. We've got ten all told.'

'Ten! What, from one litter?' I exclaimed, horrified.

'No, no o' course not, maid! Them's from two bitches!' He snorted with laughter at my not knowing such a simple thing as that.

'How are you going to find homes for all that lot?'

'Dunno, 'spose they'll go, just like they allis does, but if not,' he shrugged, 'we'll just keep um.'

That was the last thing I wanted, yet more barking dogs around the place. We'd spent many a disturbed night, when the tribe of dogs heard a fox, or something similar. It only took one of them to set the whole pack off in the yard, leading to a round robin of barking for miles around.

'Have you tried advertising them,' I asked, hopefully.

'No... us 'uve never done that. Might be worth a try though. Could you put one of them there advertisers in for us?'

'Don't see why not. I could do it this afternoon, if you like. It's probably best to stick it in Saturday's Western Morning News. You always see lots of ads for animals for sale in there. What's your telephone number, so that I can put it in the advert?'

I knew they had a telephone because you could often hear it ringing, even from our yard. It must have had some sort of amplified bell attached to it, so that you could hear it ringing outside.

'Dunno I'm sure. Couldn't you put your one on it 'cause our one don't always ring,' he said lying through his teeth.

'Well, I suppose I could, but it would be better if you had yours in the advert, so that when people wanted to see the puppies, you'd be there to show them. I shan't know when to tell people to call,' I reasoned.

'They'm just in this 'ere yard at the back, like. You can just get 'em to come and see over the wall can't you?' He pointed to some cardboard boxes stacked against the farmhouse wall, at the back of the yard. This must be where the bitches gave birth and kept their pups. Poor old things.

'All right. As soon as I get some phone calls I'll let you know.' Anything for a quiet life, I thought. And of course to get rid of some of those pups.

'You could just sell 'em if you likes,' he said thoughtfully, as if he were just about to do me a big favour.

'Could I? How much do you want for them?' And here I was thinking it would be difficult enough to get rid of them "free to a good home".

'Oooh, I dunno...a tenner would be nice.'

'Yes it might be, but you'll never get that much for them, surely? I would've thought that "free to a good home" would be your best bet.'

'Them's good dogs they are. They'd be good with sheep. I seen many a sheep dog go for fifty quid or more,' he said indignantly, absolutely aghast at my suggestion that they go free.

'Well, you could start off at ten pounds, but you'll probably only end up with a fiver, if you're lucky,' I countered.

'I'll leave it to you maid, you'll do a proper job, I'm sure.' James's confidence in me knew no bounds.

'I'll do my best with them then, but you'll have to let me have them over here, because I'm not going into your yard while you're out.'

'I'll put 'em all in a box then on Sat'd'y mornin' and put 'em over by your gate.' Clutching his squirming bundle, he shuffled off, well pleased with himself and the deal he had done.

I rang the Western Morning News that afternoon, managing to get the advert in for the next Saturday. I caught sight of the pups on quite a few occasions during the next few days. It was amusing to see them trooping up and down the track, usually following one of the adult dogs. Perhaps Mum or Dad? Little fat sassy bundles of fluff on stumpy short legs. They looked perfectly healthy and happy but it would be good to see them going to good homes, rather than joining the ever growing pack that roamed around the barns.

The phone calls started early the next Saturday morning. I was in a bit of a quandary as to what exactly to tell people regarding the pups' origins; starting off a bit hesitantly with the first couple of calls. To my own ears I sounded a bit guilty, like a dog-napper or something, especially when one lady asked me how many I had left and I had to admit to having ten! I didn't want to give people the idea that the pups belonged to me. They were nice enough to look at now but would grow into real horrors, I was sure, and didn't fancy them being brought back at a later date. Not only that, they had a very bad case of fleas and, I formed the impression from the sight of their over-extended tummies, they might also have worms. I soon got into the swing of a good sales patter, inviting people out to see the brood, as and when they wanted to. The Grimms had gone out earlier on, so I felt that I had a free rein to take offers on the pups, if it meant that they were going to good homes.

Later that afternoon I had sold eight of the pups, making a grand total of sixty pounds for my efforts. I felt very pleased with myself. The phone calls had dried up to just a trickle and no one else seemed to be interested in coming to see the remaining two. It seemed that everyone wanted dogs and didn't want to see the last two bitches. There would be time perhaps tomorrow for more phone calls.

I put the box, with the two puppies in it, over the wall, hoping that they would be put with the other dogs as soon as the Grimms got back from market. It was fair to say that the Grimms were often a little unsteady on their feet when they returned from a day out, but I had never seen them wilfully neglect their animals. They would always manage to stagger round with a bucket, lopsidedly spilling feed behind them, however inebriated they became.

True to form, when they turned up, they'd definitely had a skinful apiece. I was in the orchard and had a good view of them. George was the first one I saw, he was attempting to open the top gate. It took a fair while. After the car had finally driven through the gate, he closed it and then staggered down the lane in its wake. His hat was always set at a jaunty angle, however, today it was literally only just perched on the back of his head. I waved but should have known better. At first I thought he was waving back but as he came towards the fence nearest to me, I could hear him muttering and cursing and now see that he was actually waving his fist at me.

'Gypsies, nothing but bleddy gypsies, you lot,' he raged. 'Thieves that's all you are.'

Oh hell, he was in one of his fighting moods was he? Talk about ungrateful. Here was me, practically taking the whole day off just to sell their pups and all they could do was hurl abuse at me! They had obviously been discussing the loss of "their" barns over a few whiskies. I'm sure they probably thought we'd stolen them or, more likely, still owed them the money! I knew the rantings and ravings were all a load of booze-talk but nevertheless it always left me a bit shaky and, depending on my own mood, a bit tearful. I thought I'd best keep my head down and ignore him. He'd probably have forgotten all about it by tomorrow. I moved further into the orchard and he eventually got tired of barracking me and moved off down the lane.

The Grimms were fairly predictable; most mornings would see them up at dawn, dressed in their work clothes, looking busy and feeding animals. Depending on the weather, James might be in the garden doing a bit of weeding or deadheading the roses. Ted used to be the one that mucked out the yard, generally cleaning up after the animals, but after his death this was a job that didn't appear to be top priority. They had a fairly ancient cow, her hipbones and ribs prominent, plus her slightly plumper calf that, when the weather was inclement or the grass was sparse, mainly lived in the tumble down shippon and yard that was set at the front of the farmhouse. Their dung used to lie where it fell. Every now and then, it would grow to such momentous proportions that George would go and poke a bit of muck around, in a very desultory way, as if his heart wasn't really in it. Eventually, he might even summon the enthusiasm to pile it all in the middle of the yard, like Ted used to do. When it got so high as to be unmanageable, they would ask Gordon to give them a hand to move it. They would all then pitch in to pile the foul smelling concoction onto the trailer, after which Gordon would hitch it up to the Land Rover and drive it off to be tipped in the field. Much

much later, when it was as hard to work as it could possibly get, the Grimms would set to with pitchforks and fling it around, willy-nilly, where it would be eventually watered in and give the pasture a much-needed boost.

Normally by ten o'clock they would have disappeared inside. Half an hour later, looking spruce in their town clothes, they would be ready for the off. James would drive their white car out of the carport, with Verity already loaded, in the front seat. This "loading" was kept a very secret affair and we had only caught sight of it a couple of times, which was amazing, considering it was an event that happened almost every day. George or, when he was alive, Ted would act as a "look out" and, when they thought the coast was clear, James would very swiftly carry Verity, from the back door into the car. She would be on his back with his arms tightly clamping her legs, which were wound round his waist, her arms clutching his neck - just like a little monkey. Once the precious bundle was loaded, the two West Highland terriers would be the next to go onto the back seat. James was always the driver: George would walk in front of the car, up the lane to the gate at the top by the main road. He would then open the gate, allow the car to drive through, and then take his seat in the back, with the two Westies. Off they would go into town to do the shopping. Sometimes they went into Exeter, where they had an arrangement with one of the supermarkets to take their stale bread, which they gleefully told us they got for just one penny a loaf. We were puzzled at first seeing them unload a complete tray of bread, wondering why they needed so much, but were enlightened when we saw them hand-feeding the cattle and sheep with floppy slices of white bread.

Wednesday was a day reserved for the market at Newton Abbot. Just occasionally they would still go to the races at either Newton Abbot or Exeter, but after their heavy gambling losses that resulted in them loosing the barns to us, I think they must have learned their lesson and didn't go so often. Whatever day it was, though, and whichever town they had done their shopping in, they would always end up at some pub or other. One of their favourite routes back from the market at Newton Abbot was the Teign Valley, where there are a wealth of pubs en-route. Sometimes it would be the Travellers Rest, which was our nearest. Other customers at the Travellers Rest had already told us that they would leave Verity in the car, with the two Westies. The two men would go inside, keeping Verity topped up with whisky on a regular basis. Sometimes when they weren't quick enough in getting her a refill, she would bellow from the car 'another bleddy whisky you lot, and 'urry up about it'.

It was normal for them to get back home between three o'clock and three-thirty and we had found out the hard way, that it was best to try and avoid them at all costs. It wasn't always possible, of course, and we had both taken quite a lot of verbal abuse during the months we had lived next door. I tried my best to ignore it, because the change in their characters from sober to tanked up, was alarming and, I must admit, frightened me. They were fairly shy and reasonably polite when sober, but absolute hell when the whisky was talking. I was always too scared of them to retaliate but Gordon could never resist, either winding them up even further, or just hurling his own insults.

One afternoon while Gordon was working from the scaffolding that had been set out on the front of the barn, the inebriated Grimms returned from a session. The car drew up in front of the farmhouse, and opposite where Gordon was working, waiting for George to make his unsteady way down the lane to remove the barrier to the carport.

Not being one to waste an opportunity, James wound down the window of the car and yelled 'Yer nothin' but a load of rubbish the lot of yer. Clear off out of it and leave our barns alone.'

'What on earth are you on about now, you drunken old idiot,' yelled Gordon in return. 'You can't even walk let alone drive a car. It's a pity the police don't breathalyse you and give us all a break.'

'You oughter go back from where you come from. Taking away a man's property. It aint right, nor proper.'

'Oh go and soak yourself you stupid old bastard,' spat Gordon in disgust.

'I'm going to tell your boss about you!' James replied, absolutely purple with rage.

Quite what he meant by this statement, we were never to find out. He must have known we were building for ourselves. Maybe he meant me as the boss! What a joke...

Unfortunately the telephone calls about the pups really did dry up and the other two pups remained to build the new dynasty of ugly Grimm dogs. I handed the cash I'd collected to James the next time I saw him looking relatively sober.

'James, here's the money I've taken for the puppies. I'm sorry but the other two don't seem to have any takers for now. Do you want me to put another ad in for this weekend, or perhaps try another newspaper?' I asked him

'Oh, proper job, thanks maid.' He peered into the bag containing the cash. 'Where's the rest of the money? Did you keep back a bit of a commission?' he asked grinning.

'No, I did not, you cheeky devil. You haven't even paid me for the advert yet. You said to get what I could for them; not everyone wanted to pay ten pounds. One family wanted two pups, so I let those go for a tenner the pair. They had a lovely couple of children with them and they just fell in love with a puppy each.' I explained indignantly, feeling more than a little put upon.

'On'y jokin' you. Yes, put another of them things in the paper and us ull get us another few pounds,' he said. His disreputable face was covered in masses of white bristles, which stuck out sharply like a wiry hedgehog, and the flesh round his eyes crinkled in good humour. He was a nice old boy when he was sober.

'OK, but you'll have to pay me for the ads. Don't forget.'

'They'm free aint they?' he asked, cocking his head on one side.

'No they're not! I'll have to find out just how much they charged me for them and let you know. I paid for them on my credit card: when I get the statement, I can tell you how much they were.' What a nerve that man had. I'm sure he thought I was here to wait on him hand and foot. In his heart he thought he was still lord of all he surveyed and I was a member of his staff, just here to do his bidding.

Despite trying another advert, the pups still remained with us, growing by the day. Unfortunately this also meant that they looked less attractive by the day. Their roundness had disappeared and their snouts seemed longer and somehow sharper. We inevitable, had to resign ourselves to the fact that they would stay and swell the numbers in the Grimm pack.

A couple of months later, we were to find out that this little episode would be repeated regularly. The Grimms' dogs were very good breeders and they always appeared to have a bitch in season. I think making sixty pounds on two litters went to their heads, making them realised what a good source of pin money the puppies were. They also had a ready-made mug to do the advertising and take the phone calls. It was money for old rope in their eyes. I was, however, a willing accomplice, in a way, as finding homes for them was much better all round, then letting them add to the pack.

CHAPTER TEN
CHRISTMAS WAS COMING AND GEORGE'S
TURKEYS WERE GETTING FAT

As autumn advanced into winter our work pace, by virtue of necessity, increased to a feverish rate. No longer did we allow ourselves the luxury of even a whole Sunday off. We began to think ourselves exceedingly privileged if we managed just a couple of hours for leisure activities. We were bang on target with the work schedule, and the adrenalin rush of accomplishment was still with us.

The roof was complete, all the drainpipes fitted, fascia boards stained and the scaffolding had, at long last, been removed. Just removing the scaffolding and sweeping up the mess that lay behind it, made such a huge difference. The barn was starting to look more like a house than a building site - at least from the outside. The inside was more piecemeal. The internal walls were all in place and the plasterers were still doing a sterling job, but had not completely finished. On the upper floor, the floor boarding was completed but, so far, the doors, skirting and architraves had not even been started.

It was a brave resolution then, that at the beginning of November we made the decision to move into the barn. The weather had been cold and dismal and the caravan was getting very squalid, with wet dogs constantly in and out. Absolutely anything was going to be better than that. All we required was one bedroom and bathroom to be completed before we moved. We concentrated on these two rooms for the next few days, until we were completely satisfied. We set up our bed in the middle of the bedroom floor - just on the bare floorboards - but thought ourselves in paradise. That evening, I celebrated with a bottle of champagne, sipped leisurely whilst soaking in a foaming bubble bath surrounded by flickering candles. Oh what absolute bliss. My first bath for months. Our en-suite bathroom was a total pleasure to be in. It was spacious and, above all, warm. The radiator in that room was a bleed-off radiator that took excess heat from the Rayburn, and always seemed to be warm, whether the central heating was switched on or not. I lay until my skin shrivelled and I grew pink and heady with steam and bubbles.

We had thought that the day we moved in would remove some of the work pressure, but not so. My sister had telephoned a few weeks before to say she had booked her flights to England. The problem was that, as it was Christmas, the flights were extremely expensive the nearer to the holiday your journey took place. Her solution to this problem was to book her flights so that they could travel two weeks before Christmas! This wouldn't have been such a problem for a normal family, but for us it was an absolute tragedy. We still had skirting boards to fit to the rest of the bedrooms, the plumber had three more bathrooms to equip, there was no kitchen at all and, last but not least, no floor tiling had been started to the kitchen and utility rooms. And all this to complete in just five weeks. Piece of cake! *Ha!*

By dint of working even more hours, and cracking the whip with the plumber, we were able to at least get all the bedrooms and bathrooms habitable. There was even time to have carpet fitted to the entire upper floor and the sitting room downstairs. The ground floor, with the exception of the carpeted sitting room, unfortunately, remained bare. When they first arrived, the startled look on my American brother-in-law's face said that he didn't think much of the cardboard we had laid on the floor but, like the rest of the muddle, he soon got used to it. The barn had full central heating and, with the added benefit of an enormous wood-burning stove in the sitting room inglenook and the Rayburn in the kitchen, at least the house was warm and welcoming.

We realised very early on that it would be an impossible task to complete the kitchen before our visitors arrived. The kitchen was a huge room measuring a good thirty feet by about sixteen feet, with a large central island unit housing a Belfast sink. There was also a walk-in storeroom that would be used as a pantry. The carcassing for the units would all be made and fitted by Gordon, but we had ordered doors and cupboard fronts to be made by a local cabinetmaker, Dawson, out of reclaimed oak. Unfortunately, these would not be ready for at least another month. The worktops and flooring we had chosen were to be in black slate. All very time consuming to lay and grout. We thought that a good alternative would be to concentrate instead on the utility room. This was larger than many other kitchens I'd had in the past. There was even a storeroom to put away things such as vacuum cleaners and the like. All in all, an immense space. The carcassing for the floor and wall units were rapidly made and fitted by Gordon and slate worktops placed on top. There was no time for any fancywork – door or cupboard fronts – but at least it functioned as a make-do kitchen. I was able to get all my pots and pans unpacked and tidied away before the visitors arrived.

Gordon made the long trek to Heathrow airport to collect the American contingent, and then just two days later my parents arrived. They had flown from Alderney to Exeter so, fortunately, it was only a very short journey to collect them from Exeter airport.

In the run up to "the big day" we did manage to get one or two small jobs out of the way but, largely, had to bow to the fact that we had guests in the house, keeping work and disruption to a minimum.

The Grimms seemed to be busy dashing to and fro, very mysteriously, but at least keeping quiet and out of our way. I was terrified in case they went on a bender and started to throw insults at my mother or the children, but I needn't have worried, they had other fish to fry. Or should I have said turkeys to pluck?

A couple of days before Christmas day, a fleet of cars began to appear at the Grimms' gate. All very odd. One of the callers rang our doorbell.

'Sorry to bother you, but do you know where James is?' she asked.

'No sorry I don't. More than likely he's gone into town. Do you want to leave a message?' I asked.

'No thanks; I've just called to pick up my turkey. I'll come back later this afternoon.'

So *that's* what they'd been up to! In one of their yarning sessions, they'd told us how one of their little sidelines in the past had been to breed turkeys for Christmas and Easter, in the deep litter house, selling them at the farm gate. They certainly weren't doing that this year, though, because they were keeping their tribe of dogs in the deep litter house and there had been no sight nor sound of a turkey around the farm.

'Hey Gordon, guess what? The Grimms are selling turkeys!' I chortled.

'Selling turkeys, what are you talking about now?' he queried, raising an eyebrow.

'I've just had a woman asking if I knew where James was, as she'd come to pick up her turkey. They must be getting them from the market, or the meat auction, and passing them off as their own.'

'Typical, they're such schemers, I wouldn't put it past them,' he snorted.

Later that afternoon when I saw James, I told him that a woman had been looking for him, to pick up her turkey. He then asked me what size turkey I would like!

'James, that's really good of you to think of us but I've already got a turkey on order from the butcher in the village. It's much too late to cancel it now.'

'I been saving one special for you,' he cajoled, his teeth doing their usual shuffle while he struggled to keep them in his mouth.

'Sorry, you didn't say anything about a turkey and it's too late now. I'm picking it up on Christmas Eve and I can't let the butcher down.'

'You got plenty of folks in doors to eat up a second one haven't you? They's only little ones. No more'en twenty pounds a piece in weight.'

He hadn't mentioned where he'd got them from! 'No, really, they'll all be sick of turkey by the time they've finished,' I reasoned.

'Well, what about you have it, and put it in your freezer, then?' he coaxed 'an' if you don't eat it, well us'll buy it back from you, how about that?'

'I wouldn't have the room, James, I've only got a very small freezer.' I could see this conversation continuing ad infinitum, so made my excuses and dashed off.

The steady stream of customers for the Grimms' turkeys

only diminished late on Christmas Eve. Obviously it was big business and well worth their efforts financially.

Christmas day arrived and I was just about as disorganised, as it was possible to be. The Christmas tree trimming had been relegated to the two children, but none of the preparation for lunch had been done, and most of the presents were still not wrapped. The kitchen was full of dirty glasses and plates from the night before; not to mention breakfast to organise. I'd got up before dawn to try and get some of the clutter out of the way, with Gordon following me down soon after, helping to clean out the ashes and re-light the wood-burning stove.

He came in with an arm full of logs, together with the news that we were invited for drinks at the Grimms, for eleven o'clock sharp. I was definitely not best pleased to hear *that* little piece of information. We had a house full of guests to be entertained and, not only that, I had a full turkey roast, with all the trimmings, to cook for eight people that morning. I could well do without drinks at the Grimms that was for sure. Second thoughts, though, I badly wanted to see inside the house of horrors and at last to meet the mysterious "sister" face to face. This was something I did covet. It was our first invitation inside the house, consequently, could not be missed, whatever the sacrifice.

Armed with a bottle of whisky (which we knew full well was their favourite tipple), we set off promptly at eleven as instructed. As we walked through our yard, we saw James hovering in the garden, waiting to let us in through the little side gate. We had hardly ever seen this little gate open: it normally had a large chain and padlock wrapped round it to deter visitors. We were honoured. He waved us in with a flourish, wishing us a cheery 'Merry Christmas'. The smell of whisky on his breath was extremely strong, making it obvious his celebrations had already commenced.

As we were ushered through the massive metal studded front door, into the central hall, I was trying my best to take in everything at once, without appearing to be overly interested. The hall had once been painted white but was now peeling and flaking with dandruff-like piles in the corners of the flagstone floor. The flagstones themselves were large and looked like limestone to me. I had to be careful treading on them, as one or two of them were not as flat as they once were. There were doorways leaning in all different directions, hinting at more than a little subsidence here and there.

George emerged from one of the lopsided doorways. He was wearing a shirt, wafer thin with wear and washing, rolled up to the elbows, with an exceedingly frayed collar. His trousers were equally threadbare, having numerous holes and two matching rents over the knees. Making no concessions to Christmas, he had on his usual footwear – Wellingtons - turned over at the tops showing the grey-white linty insides.   He was also sporting his normal smell, a meaty, rotting odour. 'Come in, come in, and a warm welcome to you,' he said hastily wiping his hands down his trouser legs, before shaking hands with us both.

'Merry Christmas, George. Managed to get rid of all those turkeys then?' asked Gordon.

'All but one, and I be saving that for you boyee!' He spat lustily to the left, unerringly missing his welly and leaving a wet patch on the flagstones.

'You'd better put it in your freezer then, because our turkey's in the oven right now. I don't think, even with eight people to feed, we're going to need two turkeys today!'

Obviously everyone was on fine form and, with the whisky fumes almost visible in the air above them, like a flambéed Christmas pudding, we were ushered along the passage. The room we entered had an absolutely enormous dark wooden, almost black, dining table down the centre and, I quickly counted, twelve elaborately carved high-backed chairs. At either end of the room were two colossal glass fronted dressers, crammed with all kinds of interesting looking things. A shaft of sunlight streaked through the smeary windows, shimmering with dust motes, and highlighting the many grimy finger-marked surfaces.

In a chair by the side of a hissing mobile gas fire sat "sister" with a tartan rug round her legs. The mysterious Verity at long last. She was looking extremely flushed but it was difficult to see whether this was because she sat so close to the fire, or that she had gone mad with the rouge. She looked a little like the rag doll I had once made in school, with round patches of red on her cheeks, and bright crimson lipstick smeared vaguely round her lips. She was very obviously also wearing an old brown wig that looked as if it had been home-made out of knitting wool, not quite straight on her head.

Usually on Christmas day, I made a special effort with my outfit, always attempting to wear something new and attractive but I was particularly glad I had worn a chunky jumper, warm skirt and my boots that day, as the room was glacial. I was to be extremely glad of these boots in the next few minutes...

George walked into the room with the largest turkey I have ever seen, its claws clutched in his fist and its head dangling almost down to his feet. An off-white Westie, yelping in excitement and leaping up and down trying its best to nip off a piece of turkey, accompanied him. George gave the dog a boot with his welly to try and to get him to desist, but to no avail. Holding the turkey further aloft seemed to do the trick, although this was no mean feat, as the turkey had an extremely long neck and legs, and George was not the tallest of men. The dog soon lost interest, however, turning his attention first to Gordon and then sniffing around me. To my absolute astonishment, he then proceeded to cock his leg and eject a rancid smelling stream of yellow urine against my booted leg. As I said - a fortunate thing I decided to wear my boots. I glanced round the room, but no one seemed to have noticed, or if they had, this was something that happened on a regular basis and was best ignored.

'This 'ere little bitty one is your turkey. I'll put it in the freezer for you and you can 'ave it whenever you fancies it,' George said hanging onto the claws of the dangling turkey and doing his best to keep the head off the floor.

'You may find you want it yourselves before then but, if it's still there at Easter, you can put my name on it,' Gordon concurred, full of the Christmas spirit.

George took himself and the turkey out of the room. James, who had started to dispense the drinks, had, without too much trouble, persuaded Gordon to have a whisky. This was totally out of character, as normally Gordon would only drink beer or wine. I could see trouble brewing ahead. When it came to my turn, I was in a bit of a quandary. They had the drinks set out on a little table, not far from Verity, with two bottles of whisky, one of gin and a few mixers. Not being too keen on either of these, I asked if they had wine. No, but they did have sherry or port. Port it was. The port came in a glass smeared with the ghost of a thousand lips, cloudy all around the rim. The surface of the drink seemed to have a slick on the surface, with bits of something unidentifiable floating in it. *What on earth was it?* My stomach gave a great heave, as I manfully took a tentative sip. I was trying my hardest not to think of the germs I could be in contact with; striving not to imagine the havoc a jippy tummy would wreak over the coming days. Without too many alternatives at my disposal, I quickly came to the conclusion that the alcohol would probably give me at least a little protection.

'Aah, look 'ere, maid, I want to show you these 'ere joanies,' said James. He had talked of "joanies" before and I had never quite worked out exactly what joanies were but thought they might be old jugs or nick-knacks. He knew I collected jugs myself from our previous conversations and had told me that he had lots of "joanies" to show me sometime. He seemed to have the erroneous impression that I knew something about antiques and had asked me if I would look at his collection and give him some idea of what it was worth. The time had come to put my limited knowledge to the test.

He handed me a little painted plaster model. It was the sort of thing my sister and I had made as children from a kit. You know the sort - where you get a dark pink mould to pour Plaster of Paris into? Then, when it had set, you had to wrestle and heave, finally managing to pull the pink bit off, you could then paint it in various lurid colours? This was a pixie; about four inches high, a bit faded now and a little chipped round the edges, but definitely looking hand-made by some child or other.

'This is me,' he said his chest puffing up 'and this 'ere is our George!' He handed me another plaster model of a little boy, equally faded and chipped.

Without giving me time to find a suitably tactful comment, he whisked it out of my hand and gave me another model of a little girl. 'Let me guess, this must be Verity?'

"Tis too! 'Aint they nice? They'm proper, is them.' He grinned, as proud as punch.

'Yes, they're lovely. Worth hanging onto, I should say, they might be worth a lot of money one day,' I said diplomatically.

Glancing around the room I could see that most of their "joanies" were similar to the plaster models he had already shown me, however, there were one or two surprisingly nice things. The two dressers were bulging with bits and pieces and I could see they had some pretty china cups and saucers. Nestling among them was a six-inch pink plastic naked doll. *Best not to ask.*

Verity didn't join in the general conversation, instead, sitting nodding and, occasionally, making grunty noises of agreement. She did, however, become animated when George produced some old photographs, especially when he handed me a large photograph of a young woman of about twenty. The person was unrecognisable now but George informed me, breathing over my shoulder, that it was Verity in her youth. The photograph showed only the upper body: she was dressed in a green silk evening gown, her lovely curly dark brown hair and clear complexion were a joy to look at. Verity became quite maudlin gazing at all the old photographs, but was soon back in the spirit of things, when more drinks were handed round.

I was, fortunately, able to get away with only drinking a sip or two of my drink. I would let James fill the glass up again, so that it looked as if I had had quite a few refills, without actually drinking very much at all. This ploy either hadn't occurred to Gordon, or he was just enjoying a glass of whisky for a change. Quite a few rounds had been dispensed before I started to make general "we must go soon," hints. Neither Gordon or the Grimms took a bit of notice of my hints and, inevitably, another bottle of whisky was produced from under the table, to replace the two empty ones. Yet another round was dispensed. I could see from the large ornate mantle clock, that the time was now getting on for one o'clock and I could almost smell my turkey burning from where I was sitting. My agitation started to gather momentum. I thought the Grimms would be glad to get rid of us too, as they'd said that they were intending to eat at one o'clock. I hadn't seen much sign of them going in and out, to the kitchen, to attend pans etc. Nor for that matter could I smell anything at all cooking.

'Sorry folks but we must go, otherwise I'll not be in any fit state to cook the lunch,' I eventually said.
Gordon didn't seem to be too keen to end the party and said, 'Yes, off you go. I'll just have one more for the road and join you later.'

I shook hands all round, kissed Verity on her rosy cheek, and thanked them for their hospitality. James showed me out of the front door, and I gratefully made my escape.

When I got back into my own kitchen, my sister and mother had rallied round and had been basting and peeling for the last couple of hours. They had all been waiting for us to arrive back, before anyone opened their Christmas presents, and the poor kids were sat glumly gazing at their unopened parcels. Taking pity on them, I suggested they open a couple of theirs immediately, and then we could do the rest when Gordon came back. Was I being optimistic at this point, or ignoring all the warning signs? My husband loved a good party and had never, to my knowledge, been anything but the last to leave. Surely, though, with the family here and the smell of his Christmas dinner wafting over to him, he wouldn't keep us waiting too long? Would he? *Oh yes he would!*

I kept dashing anxiously between the kitchen and the little square window in the front door - the only one that overlooked the Grimms' house - to see if I could see any signs of activity. Nothing at all. Not a sausage. Three o'clock arrived and I had just about decided that I'd had enough and was going to haul him ignominiously out of there, when I saw the front door open. George came out with Gordon following behind. Both of them stood, or should I say, swayed, on the doorstep with their arms around each other's shoulders. Spiritual friends. They were both wearing paper hats! Lots of backslapping went on and then Gordon set off, weaving down the path, to the front gate. It took him a long time to get it open but he eventually managed it, disappearing from my sight. George stood on the front step for a while, waving merrily, with his paper hat askew. He then suddenly lurched to one side; sitting down heavily on the ground. He stayed slumped on the floor, with his back against the doorjamb, his head lolling on his shoulder.

Gordon had by this time arrived in the kitchen, so I hurried through to find out what had happened, and to serve, at long last, our very over-cooked lunch. He looked quite sensible, and it was only when he tried to talk that you could tell that he was definitely not sober.

'George is still on the front doorstep and I don't think he can get up. I hope he hasn't hurt himself,' was my first comment.

'You should see the others! When I left, Verity was slumped in the chair with her wig on the floor, and James was lying flat on his back, snoring. He's the one meant to be cooking the lunch!'

'Uncle Gordon, Uncle Gordon, come with us, we want to open all the presents,' two excited little boys chorused, fortunately timely cutting short any uncharitable, or un-Christmas-like comments, I could have made. So off Uncle Gordon went and our Christmas day continued. When, later, I had time to look, George had disappeared off the doorstep but I didn't somehow think either of the Grimms would manage to cook their Christmas lunch until the next day at the very earliest.

When we had all recovered from the gargantuan meal, Joe my brother-in-law was elected to wash up with Gordon, now a little more sober, wiping and putting away. I was still clearing the table, had just put a pile of plates next to the sink while Joe, who had already run the hot water, was standing squeezing the washing up liquid into the sudsy water. The only problem was that he didn't stop. The thick green liquid just kept pouring out of the bottle. Just as the level of the soap suds reached the top of the working surface, soap bubbles floating in the air, I thought it was about time I intervened, 'Erm...Joe, I think you've probably got enough soap in there now.'

'Oh! Yes, I think you're right!' he replied. 'We've got a dish washer at home so I'm not used to this.' He was swivelling the long arm of the mixer tap backwards and forwards with a puzzled expression on his face. 'Is there a problem with your faucet?' he asked 'It looks like it could sure do with some attention to me.'

'No, Joe, it's designed to do that. Haven't you seen a tap like that before?'

'No, I just thought it was loose and needed tightening up!' Poor Joe by this time was up to his armpits in suds. They were fast escaping over the rim of the sink, to float gently away down the face of the cupboard door, and onto the floor.

The subject of plumbing came up again, when I was in the middle of cooking lunch, on Boxing Day. My sister came into the kitchen to find me. 'We've got a bit of a problem with the plumbing in our bathroom,' she announced, dramatically.

My heart sank like a stone at the implications behind this proclamation. We had been in such a rush to complete all the bathrooms before our visitors arrived, that my over-active imagination had already seen a gargantuan leak, the size of Niagara Falls, spurting from under the floorboards. Converting old buildings is always rife with problems, but plumbing always seems to give more aggravation and inconvenience than any other. To put any major plumbing blunder right would more than likely involve days of disruption, with no water on tap, and toilets out of action. I just couldn't imagine any of our visitors using an empty paint tin - something we had resorted to in the past - when a toilet had been out of action for a whole bank holiday weekend, until we could obtain a vital coupling for a pipe.

The barn was well equipped with bathrooms, so any unidentified hiccup might well be a major disaster. Upstairs there were two en-suites, one family bathroom and - on the ground floor - a cloakroom complete with toilet. This number of bathrooms might seem like overkill, with only two regular occupants, but with a house full of guests - totally essential.

'What's the problem?' I said, nonchalantly, still stirring and chopping, doing my best not to let my concern show. 'Is there a leak?'

She looked a little embarrassed, 'No...it's more of a blockage. I'm afraid we've got a floater.'

'A what?' I said, not sure I'd heard her correctly. She often used American terminology, after living in the United States of America for fifteen years, that I wasn't entirely familiar with.

'A floater...you know...a submarine!' she explained.

'Oh!' I said, penny dropping at long last. I started to laugh then, more with relief, that it was only a minor inconvenience and not the major disaster my imagination had conjured up.

'It's no laughing matter,' she said, indignantly. 'It just won't flush away. I've tried dozens of times and even taken a bucket of water up.'

'Oh stop worrying about it. It'll go away in it's own good time. Just leave it be,' was my solution.

'I can't do that,' she replied, firmly, 'someone else might want to use the bathroom.'

'You'll just have to own up and tell 'em it was you!' I replied under the impression that her embarrassment was on her own behalf.

'It wasn't me!' she said indignantly 'it was Matthew.'
Matthew was my lovely ten-year-old nephew. Very sweet but obviously he had a bowel problem, by the sound of it.

'I don't know what you expect me to do about it,' I said, less than sympathetically. I'd been hard at it for days, cooking and cleaning, and certainly didn't feel that unblocking toilets was a task I wanted to take on, just at that moment in time.

'Just come and look at it...please,' she pleaded.

'Oh, all right, just a second while I turn this pan down.'

We trooped up the stairs to their bathroom, to inspect the offending article. 'Good grief!' It was an absolute monster. 'How on earth could a ten-year-old produce something of that size?' I gasped. By this time I was hardly able to stand. I was leaning against the wall, quietly shaking with laughter. 'Are you sure it's not yours, or maybe Joe's? He's a big chap when all's said and done!' I was teasing her now.

'It's all right for you to laugh: you haven't got children. It happens all the time in our house.' She was a little on the huffy side by now. 'The toilet was blocked so badly once, that it overflowed everywhere, and I've always been in a panic that it could happen any time again. I'm always on tenterhooks until it's cleared.'

This thought sobered me up somewhat, and I tried very hard to pull myself together. 'Just don't worry about it, leave it and use one of the other toilets for now. I'll ask Gordon to deal with it later.'

'No I can't,' she insisted, 'I've got to deal with it now.'

'Well, what do you do when it happens at home then?' I asked, curiously.

'If they're that size,' she said gesturing towards the pan, 'I cut them up into small bits.'

'You are joking me, aren't you?' Her look said, "No she wasn't". 'What with, for goodness sake? A knife and fork?'

'It's all very well to laugh, but this is serious. As it happens, I do - plastic ones! Have you got any?'

It was some time before I could answer her. 'Yes,' I spluttered, 'I think there's some in one of the drawers downstairs.'

With that she went in search of her cutlery to cut the offending substance into bite size pieces.

CHAPTER ELEVEN
A BREAKAGE OR TWO

We were back to normal, our visitors having long since departed, when a deafening crashing noise interrupted our well-earned tea break. The noise sounded as if a whole building had fallen down, somewhere in the region of the lane, out at the front of the house.

'What on earth was that,' I asked startled, slopping my tea all over the table.

'God knows, but it sounded like something serious coming down,' replied Gordon.

We both shot round to the front of the house, not knowing what to expect. Literally the dust was still settling on a pile of bricks and rubble, which had once been one of the large ornate chimneys belonging to the farmhouse. James and George, looking extremely dishevelled, and covered in a film of grey dust, emerged from their backdoor, just the other side of the huge pile of rubble.

'Are you all OK?' we both asked anxiously, almost in unison.

'Yes, we'em all proper, but our chimeney's 'ad it,' replied George, ruefully scratching his head.

'I'm buggered if I know what's 'appened to it.' said James.

It was fairly easy to understand how it had happened, even to a novice builder like me. The house was in such a run down state, that it seemed impossible for any maintenance to have been carried out on it for many years. Walls were bulging and plaster peeling: subsidence evident in every leaning door and window. The wooden window frames were a delicate shade of pink; all the paint appeared to have worn off and the original base coating now showed through. It was entirely possible that they had never seen a coat of paint at all. Many of the panes of glass were either missing or cracked; someone had even carried out a running repair with cardboard. The temporary patch had been there as long as we had and, most probably, for many years before that. Occasionally, in a heavy wind, the roof would lose a few slates, but these were never replaced. With no felt under the slates, this must have made the bedrooms, beneath the roof, very uncomfortable indeed. I could well imagine the dozens of bowls and buckets set out around the beds to catch the drips, whenever it rained.

'Is there any damage inside the house, or is it just this mess out here,' asked Gordon, gesturing to the pile in front of us.

'It's a mite dusty, but the wall is still standing. All them there chimeny bricks is out here,' answered James.

'That's good, but you'll have to get it seen to soon, otherwise the rain will get into the outside walls and only make things worse,' advised Gordon.

The chimney had almost peeled itself away from the cob wall of the house, leaving a raw patch of exposed cob. Any rain or damp that managed to penetrate a cob wall was a disaster. It seemed that at Castle Grimm some wet had been allowed to get between the chimney, which was built mainly of brick, and the adjoining cob wall. The disaster had obviously just happened.

'Looks like a big job to me. It's a large chimney,' said Gordon reflectively, 'and it will cost quite a few thousand to put right, I should think.'

''Ow about you giving us a hand doin' it?' George asked, his roguish grin still not failing him.

'I should think it's more a job for the insurance, if you ask me,' replied Gordon.

George sucked his teeth and shook his head. 'Us ul 'ave to give 'em a ring and see,' he said splitting a gob to the left of him.

Were they even insured? It seemed a good bet, to both Gordon and I, that they more than likely weren't. Especially when the chimney remained on the floor for many weeks and the exposed cob got wetter and wetter. Cob being a mixture of clay, straw and probably horsehair, when dry was extremely strong and durable but - only when dry. If it should get wet, then it reverts to its natural state – mud. We were expecting the whole side of the house to fall down any day.

We had a visitor ourselves, at that time, so thoughts of the Grimms' problems went totally out of our heads. When our planning permission for the barn conversion was granted, one of the conditions attached to it was that the cider press remain as close to its original position as practical. At the time we congratulated ourselves on our good fortune, in that the condition wasn't more specific. At the pre-planning meeting, the Planning Officer had intimated that she would insist that the cider press be replaced in *exactly* the position she had originally seen it in. As this was immediately in front of a window opening, this would have been very bad news for us, as it would have severely restricted the light at that end of the sitting room. The openings were not plentiful, nor large, so any obscuring of the light was to be avoided like the plague.

We had decided to stretch the planning condition to its fullest extent and site the cider press outside, in the courtyard, where it would enhance the garden. There was a tremendous amount of space in the sitting room but the press, being in a pretty poor state, just didn't add anything to the interior décor. It was also quite an impressive piece of machinery, being about eight feet tall and ten wide. It had seemed a pity, in a way, for it to be hidden inside the house, where only we would see it. It was old enough to be a museum piece and we had in fact already offered it to a museum on Dartmoor, which was dedicated to farming. They had declined our offer for some reason or other, which was just as well, in the circumstances. We had settled, in the end, that the best place for it would be to position it in the courtyard, where it could be seen and admired, not only from our lane, but also from the main road, which ran along the top of our land.

Our visitor was the Enforcement Officer from the local Planning Department. Apparently a neighbour had advised them that the cider press was now situated outside the house, and not where the planners had intended it to be. The

Enforcement Officer insisted that the press be reinstated inside the barn, as close to the position we had found it in as possible - within the week. I can't think what would have happened if the museum had accepted our very generous donation! As soon as the Enforcement Office had left, we both looked at each other in dismayed silence for a few seconds.

'What next?' Gordon eventually said, 'I suppose we've got the Grimms to thank for that visit. I know they've got a copy of our planning permission, because George waved it in my face the other day, so he's probably going to check on every last little detail.' He heaved a great sigh. 'Oh well, we've got no alternative but to get the press cleaned up and put back inside, otherwise we may find ourselves in trouble.'

He was still thinking about the situation a few hours later. ' I've got a good mind to shop our dear neighbours to the listed building people, before their house finally falls down,' he fumed. 'It would serve them right to have a taste of their own medicine for a change.'

The Grimms' farmhouse was Grade II listed, built in the sixteenth century, and apparently quite an important building, by local standards. It was almost a small manor house and, presumably, the landowner who had originally had it built had been reasonably well–to-do. I knew that if you have the "honour" of owning a Grade II listed property, then it had to be maintained to a fairly high standard and these conditions were usually rigidly enforced. We had never owned a listed property, so had no experience to go on, but it seemed from the Grimms' situation that listed properties were not inspected very often. The farmhouse had obviously been in a pretty poor state for years and, if anyone had bothered to check, then I'm sure they would have been horrified at the state it was in now.

'What's the point in telling tales on them? It won't help us any. Please...just leave it, I don't want any more trouble with the Grimms,' I pleaded, trying to avert yet another row. 'In any case how do you know it was them?'

'Who else is it likely to be? No one else could be as two-faced as they are for a start.'

'I still don't want any more rows,' I reiterated firmly. 'They've been reasonably quiet lately and I'd prefer it to stay that way.'

'Have it your own way, but they really are getting my back up with all their shenanigans,' retorted Gordon.

The cider press really was in a hideous state. The metal parts were rusty; the wooden ones were liberally decorated with worm holes, presumably woodworm, although they were a little on the large side so could conceivably have been death-watch beetle, and one of the enormous crushing beams was completely missing. It was going to take countless hours of hard labour to make it presentable.

A good friend of ours, Clive, who had a metal workshop, took pity on us and removed the moving metal parts that consisted of huge cogs and metal ratchets. It took him many hours to patiently get rid of all the rust, and restore them to their former, well-oiled selves. A complete new wooden beam had to be ordered for the bottom section. The original beam measured eight feet by three feet six inches by two feet, which was not only extremely heavy but also prohibitively expensive to replace.

We had to hire a forklift truck to manoeuvre the beams back into the sitting room. It was very fortunate that one of the openings in that room was large enough for a forklift to slide the beams through the window. It was also providential that we had decided to leave the ceiling full height in this room; otherwise we would have been in serious trouble!

I stood and watched while a block and tackle were set up on the huge ceiling beam overhead, to haul the equally massive blocks of wood into position. It was nail-biting stuff but the task was accomplished without mishap. We certainly would not have been able to carry out this operation without the expert assistance of two very good friends - Clive and Henry. When the cider press was finally reassembled, it looked totally glorious, and did add that little touch of je ne sais quoi.

While Clive had been helping put the press back together, he had asked us if we wanted a brace of pheasants. He loved to go shooting in season and had bagged a couple of brace that weekend. They were actually still in the back of his 4x4 and he had already got them out before we could say nay. They were fabulously plump birds and I'm sure, if someone had already prepared them for me, then I probably wouldn't have demurred. 'Thanks Clive, but no thanks! I just wouldn't know how to go about plucking them, and the thought of all that blood-and-guts puts me off a bit.' I said.

'Oh, that's a pity. They would have been nice table birds,' he replied.

'Sorry! I'm a bit squeamish. I can always remember buying a chicken from a market, up north somewhere, and when I got it home it had all its innards in. All I managed to do was gouge a bit of the breast meat off and then chuck the rest into the bin. I'm totally useless.' I said shrugging my shoulders.

'Don't worry; I'm sure there'll be someone at the Travellers Rest who'll fancy them. That's if I don't forget they're in the car!' Clive replied.

Later on when they had all departed, George hailed me from the gate. 'What did you do with them there pheasants then?' he asked. He must have been behind the hedge, listening to the conversation, while I was taking to Clive.

'Clive took them home with him, why?' I replied.

'We love a bit of that old pheasant. Our Verity do love it.'

'Oh, sorry. I didn't think. Next time I see Clive, I'll ask him if I can have a brace for you, when he goes shooting again.'

'Proper job, maid,' said George, waving a hand, as he turned to wander off.

The cider press now firmly out of the way, our next task was to complete the kitchen. Dawson, the cabinetmaker, had at last delivered all the doors and drawer fronts and these had been fitted to the carcassing that Gordon had constructed. My job was to gently rub the lovely grained oak smooth with sandpaper, then with fine grade wire wool and, finally, to give it a few coats of glutinous bees wax. The doors and drawers were then fitted with hand-made cast-iron handles. The whole effect looked fabulous. By the time we had finished laying the blue-black Spanish slate floor, we were well pleased with our efforts.

Many weeks went by before any work commenced on the Grimms' farmhouse. In fact we had another horrendous storm before anything at all was done. This time the Grimms lost dozens of slates from the roof, which left a large gaping hole. The hole was obviously of such a size that something had to be done, and pretty fast. We saw what could have been an insurance assessor examining the damage, clipboard in hand, busily making notes. It was not long after, when a gang of six men arrived. Scaffolding was erected and work quickly got underway. The roof was repaired first, and then they laboured for days removing the old rubble from the fallen chimney. They didn't appear to put in any new foundation, but started building from the exact place the original chimney had fallen from. They started to build a massive chimney, far larger than the original for some unknown reason. The chimney was duly completed, looking rather incongruously new against the old unpainted walls but, nonetheless, the Grimms were well pleased with the job.

We were asleep one night, a few weeks later, when a crash woke us up with a start.

'What the hell was that?' yelled Gordon, coming to with a jerk.

'It was outside, I think. Sounded as if it came from the front.' I said anxiously, hoping-against-hope that it was nothing to do with us.

Gordon quickly got dressed and rushed downstairs, gathering a torch as he went out. When he came back, absolutely freezing cold, he said, 'It's the Grimms' new chimney. I can't believe it; it's fallen down again! There are just masses of bricks all over the place. Some have even fallen in the lane,' he said, shaking his head.

'At least it's not us.' I said with some relief.

'Amen to that,' he said shivering and crawling into bed. 'I'm freezing...warm me up woman!'

How or why the chimney came down again, was never explained, but it appeared that building such a large chimney from the original, probably faulty, foundations was a mistake. It took the contractors quite a few hours to clear up all the rubble next morning. This time they dug out a new footing, and the chimney was rebuilt from the ground, off this new and more solid foundation.

The gales, which had probably brought the Grimms' chimney down in the first place, gave way to heavy frosts. As the lane was on quite a steep slope, getting out of the car to open the gate, before driving onto the main road, became a seriously dicey business. I was very careful to wear sensible shoes with grips on the bottoms, and to cautiously make my way from the car, holding on to the open door, to get to the gate. Unfortunately, Kathleen, the lady who lived in the smallholding further up the lane, wasn't so lucky. Very early one morning, before we were even out of bed, someone was hammering on the kitchen door. It sounded urgent; so I went down to open the door in my nightie.

'Kathleen, whatever have you done?' I asked.

Kathleen stood trembling on the doorstep, clutching her

right wrist and looking extremely green round the gills. 'I've just fallen on the ice,' she gasped miserably.

'Come in...come in and sit down,' I said, panicking quietly inside.

'I think it must be broken. I can't move it and it hurts like hell.' By now she was sobbing in little gasps, obviously very shocked.

'Just sit tight for a minute: I'm going to get dressed and grab a coat. I'll take you straight to casualty. It'll probably be quicker than getting an ambulance out here.'

After quickly donning some warm clothes, I found a soft woollen scarf and knotted this round her neck, to give the wrist some support. I like to think I'm someone who can be relied on in an emergency but, this time, I was dithering around like a jelly. Kathleen, on the other hand, although in serious pain, was holding her own like a good 'un.

She had left her car at the top of the lane, the engine running and the driver's door wide open. She had also left the top gate swinging on its hinges - a heinous crime in the Grimms' book. But, let's face it, if you'd just had a nasty fall on ice, and think you've broken something, then the last thing you'd be thinking of is closing a gate. Unfortunately this gate had caused endless rows between the Grimms and everyone else who had the misfortune to use it. They had always insisted that it be kept closed, no matter how long you were going to be. I used to bake the apple pies for the Travellers Rest pub and, when I delivered them would sometimes, very naughty I know, not bother to close the gate behind me. There would always be a furious Grimm waiting for me when I got back, even when I'd only been gone ten minutes at most. The Grimms often allowed some of their cattle to graze on the small triangle of grass right up by the side of the gate, so no doubt having the gate closed was important to them, as there was no cattle grid. It didn't make a whole lot of sense though, when you could see that the animals were in their yard - especially if you were in a tearing hurry. Opening and closing that gate became very monotonous, and even though I was often tempted not to bother, usually complied with the Grimms' law, just for a quiet life.

Problems started years before, when the two bungalows were first built. Previously, the Luscombes had had this little lane, and the gate, to themselves but, when the bungalows were sold, the new owners also had access along the farm lane and used it daily. The gate very soon became a bone of contention. According to Kathleen, often when she returned home after dark she would find the handle of the gate smeared with excrement. Another little surprise that had lain in wait for her was a row of concrete blocks, deliberately left in the middle of the lane. Fortunately, her headlights had picked out the obstruction before she crashed into them. Kathleen had also told us that they occasionally allowed the old bull to graze up by the gate. She said that getting in and out of the car to open the gate had taken real nerve and, although she had complained to the police on more than one occasion, it seemed they were unable to do anything about the Grimms' behaviour. The Luscombes had never, even in the early days, been on speaking terms with either of the owners of the two bungalows - only hurling abuse terms.

While I was hurriedly getting myself dressed, Gordon got the Land Rover started for me, then I was able to get Kathleen settled as comfortably as possible in the passenger seat. Gordon had gone up to the top of the lane to move Kathleen's car: with its door swinging wide, it was blocking the entrance to the gate. After Gordon had driven Kathleen's car away, I started up the lane, trying to steer clear of the sheets of ice that lay on it. I was just about to drive through the open gate, when George appeared at my window, gesturing wildly. I wound the window down. 'What do you want?' I demanded not pleased at being delayed.

'That there gate's been left open again. That there bleddy 'oman knows it's to be kept closed. She needs a good beating she does,' he ranted, spitting and splothering in his fury. He had one of his hands inside the window, clutching the top of the glass, and was just raising his stick in his other hand: I thought he was about to hit me.

'What are you talking about you silly little man? Get out of my way before I run over you,' I cried, almost hysterical by this time. I didn't wait for a reply: gunned the engine and drove on. I could see him through my rear view mirror, shaking his stick with one hand and his fist with the other. Kathleen hadn't lost her sense of humour and we were both grinning stupidly after that skirmish.

'You are plucky,' she said. 'I couldn't have stood up to him like that. He makes mincemeat out of me.'

'I think it's the adrenalin helping,' I said, ruefully, 'I certainly don't think I'd be as brave as you are, if I'd just done that to my wrist. You look very pale, are you all right?'

'Just about hanging on. I do feel a bit sick ... I think I'll just put my head back for a bit, if you don't mind,' she said.
It was no more than six or so miles and I got Kathleen to casualty in record time. After an x-ray, they found it was the worst possible scenario, she'd broken her wrist in a couple of places, and there were bits of bone everywhere. They decided to operate, as it would need pinning, therefore, I had to leave her there and phone her husband to tell him to bad tidings.

When I got home, I regaled Gordon with the tale of my row with George, and then the details of Kathleen's injuries. I must admit I had forgotten to choose my words carefully and told it exactly how it was. Gordon always got so angry when any of the Grimms shouted or swore at me, that I had learned to tone things down, or even not tell him at all. Predictably, he was livid, and I could see trouble simmering away for when he next encountered either of them. I could only hope that it would all be forgotten and that the Grimms would keep their heads down for a while.

When I next saw George a few days later, he said looking sheepish, 'I'm sorry I spoke a bit sharp to 'e the other day, maid.' Obviously Gordon had already spoken to him!

'I should think so too,' I said. 'If you were lurking up by the gate, you must have seen Kathleen fall, so why didn't you help her instead of shouting at her like that?'

'I did see 'er fall,' he admitted. 'That 'oman's no good and I wouldn't lift a finger to 'elp 'er if she was dying,' he said passionately.

'Well that's an awful thing to say! You should be ashamed of yourself,' I replied quite shocked at his died-in-the-wool attitude.

'I don't care! Them's naught but trouble.'

'You disappoint me George. I didn't think you were so uncharitable.'

'I be a good Christian, I be.' he said indignantly.

Obviously we would never in a month of Sundays see eye to eye on that score.

The saga of the gate seemed to be a never-ending one. The cow and calf that had been here since we first arrived, had been housed either in the shippon in front of the farmhouse, or in the field just behind it. The calf was a bull calf and, eventually, despite being fed largely on white sliced bread, vegetable peelings, and a very small amount of hay, got to be quite a hefty muscular animal. The cow then had another calf, a female this time, which joined her brother and mother in the meagre grazing.

Every now and then, when the grass up by the gate got to be a respectable level, the Grimms would leave the cow and her two, now adult, offspring to graze it. The temper of the young bull was very unpredictable: even the Grimms went so far as to admit that he was "a nasty bugger". George told us that the bull had given him a bit of a turn when he was mucking out the yard one day. The young bull had cornered him up against the barn, and it was only the fact that George had a pitchfork in his hand that had saved his bacon. After prodding the enraged bull to keep it at bay, he had managed to jump onto the wall surrounding the yard, and thus escaped injury or even possibly death.

Most of the time the three animals would be watched over by George, while they grazed at the top of the lane. He would poke and prod the dry earth around the grassy knoll with his stick, looking nonchalantly around him, often as not accompanied by one or two of their mangy dogs. He would stand there patiently for hours, while the animals chewed at the scrubby grass.

When we first saw the bull wandering around, completely untethered and unsupervised, by the gate, we anticipated trouble. It didn't take long in coming. First one of our neighbours arrived to tell us to do something about it, and then another stopped at the gate to say almost the same thing. When the Luscombes owned the lane, both neighbours accepted that the Luscombes could almost do as they liked but, now that we owned the lane and gate, they looked to us to sort out any problems of that sort. It certainly gave my heart a bit of a flutter, seeing the bull wandering around without his chaperone, and I used to pray, on returning home from town, that it was not right up against the gate. The further away from the gate it had wandered, gave me a sporting chance of getting in and out of the car with my bits intact.

A couple of hours later, we could see that the bull was still loose in the lane and had, by this time, wandered down to stand at the gate, leading into the yard and shippon, where he was housed during the worst of the winter weather. His mother and sibling were inside the yard and he evidently wanted to go in too.

Gordon said, 'I suppose I'd better go and see if I can rouse one of the Grimms to move old bully boy back into the field or the yard.'

'Well be careful,' I warned. 'Go out of the front door and you'll avoid the bull. I think he's just outside our gate.' The gate was about twenty yards further down the lane than the front door.

Returning ten minutes later he said, 'Bloody Grimms! Of course, no one's answering the door. They must know why I want to see them and are hiding from me. They couldn't have failed to hear me knocking on the back door. They're definitely in there...I mean where else would they be for goodness sake?' *Oh dear, he was getting riled.*

'Leave it a few minutes, they'll probably go up and put the bull in the field themselves soon anyway, it's getting dark. It's certainly not worth getting yourself all worked up for: your blood pressure must be sky-high.' I said.

'Yes, you're probably right,' he agreed.

The bull was still wandering in the lane when we went to bed but I could see him safely up in the field, chomping away at the tussocks of grass, early the next morning. The Grimms had obviously been around early, although they were now nowhere to be seen.

A few days later, the situation remained unchanged. The bull was still grazing unchecked during daylight hours, free to wander around wherever he chose. His mother and sister accompanied him now. Whoever had to drive out of the gates, ran the gauntlet of all three animals. Gordon again couldn't get the Grimms to answer the door to his repeated knockings. He was completely infuriated by the whole situation. I saw him grab hold of the large yard broom, confidently striding up the lane to stand in front of the bull. He started to wave his arms around and shout, 'gerrup there...gerrup there,' which looked at first as if it was going to have the desired effect. The cow started slowly to move off towards the field, with her female counterpart, following closely behind. The bull curiously raised his head to watch but didn't attempt to follow. Gordon started to wave the broom up and down; continuing to shout 'gerrup'. Suddenly, kicking his back hooves in the air, the bull gave a great bellow and ran to join his mother and sister. Gordon followed slowly, closing the gate to the paddock behind them. Thank the Lord for that. My heart had been in my mouth as soon as I saw the bull cantering around. I thought I was a widow for sure.

We were together in the car returning from a trip into town when we next spotted the Grimms, just a few days after this incident.

'I want a word with you George,' said Gordon ominously.

'Oh ahh, so what's up then?' George asked looking as if butter wouldn't melt, and spitting his usual gob onto the path.

'Your bull's been left to roam in the lane. You told me yourself that he's dangerous so he shouldn't be allowed to wander around on his own.'

'Us was keepin' an eye on him through the window. He's no trouble,' retorted George.

'You know that's not true! In any case, I've told you before, he needs to be supervised if he's up here by the gate. You need to be standing up here with him, not inside looking out of the window. The girls don't like getting out of their cars to open the gate, when he's on his own, and I don't blame them: he's a big mean-looking animal.'

'Well I say 'e aint no trouble. I don't know what you'm so bothered about,' George said emphatically.

'Just stay with him, if he's loose, then there will be no trouble.'

'Us 'ave allis kept a bull on that there bit of grass. We've never 'ad no bother until all them gypsies moved in.'

'You're an argumentative old devil, George. I haven't got time for all this now. You might have kept the bull there in the past, but you're certainly not going to do it now. I own the gate and the lane so unless you want me to take the gate off completely then just do as I ask, there's a good chap.'

George had gone very red in the face and had started to splutter. 'You've got no right to go and say things like that. You'm nothing but a mushroom! Come up out of nowhere! It aint your lane and that there gate stays where it is,' he ordered.

My heart sank. Oh hell, here we go again. It seemed almost impossible to have a conversation with the Grimms these days without having an argument too.

'This is the last time I'm going to say this, George, if either of the neighbours have to come complaining to me again about the bull, I'm going to take that gate off, and have done with it. I'm just about up to here with the lot of you,' warned Gordon.

'And *I've* told *you*, we'll do as we please. We know a man who'll bury you six feet deep, and that'll settle you, me boyee,' George muttered ominously.

'Well it certainly wont be you!' Gordon retorted.

With that heart stopper, we left him, still spluttering and waving his stick at us.

All was quiet for a few weeks and then coming home from town I saw the bull grazing on the hillock, just the other side of the gate. Trouble. I guessed it wouldn't be too long before one of the neighbours drove through the gate and observed what I had. Kathleen stopped at our gate and rang the doorbell. Her arm was still in a sling, supporting her wrist in heavy plaster.

'Gordon, that bloody bull is loose again,' she said.

'Oh God, not again!' Gordon groaned. 'Kathleen, I have told them endless times: they just wont get it through their thick heads that I mean business. I think I'm going to have to carry out my threat to get rid of the gate,' replied Gordon.

'Why don't you? It's a pain in the bum opening and closing it: it's caused nothing but trouble for years anyway,' retorted Kathleen.

Sighing and shaking his head Gordon could do nothing but agree with her.

He left the house a few minutes later. The noise of the ensuing altercation was loud enough for me to hear from where I was standing in the kitchen. As I went out into the yard, I could see Gordon marching purposefully up the lane towards the main road. He was pretty obvious he was in a blinding rage, but I was not prepared for exactly how much of a rage he was in. When he got to the gate, he actually lifted it off it's hinges, arms bulging, like the Incredible Hulk, swung it over his head and marched across the main road with it. When he got to the other side, he bodily threw it over the hedge, where it landed in the field beyond with a thud that I could hear from where I was still standing. That gate was a five-barred gate made of iron tubing and must have measured a good nine feet across. How much it weighed was difficult to say, but heavy it must certainly have been. I knew Gordon was very strong, but didn't think he was *that* strong. He strode back down the lane and into the yard, absolutely incandescent with anger.

'If I don't deck that bloody man before the days out, it'll be a miracle,' he spat.

'What on earth did he say to make you do that? It's a wonder you didn't give yourself a hernia, lifting the gate over your head!'

'He just gave me a load of garbled rubbish as usual, about the bull being OK on it's own. Then he started poking me in the chest with his stick.'

'Oh no...' I groaned.

'I have warned him, you can't say I haven't.'

'They're never going to let is rest now. You know that gate's always been a problem. It's almost a symbol to them. As if they're telling the rest of the world while the gate's attached, they're still in charge,' I brooded.

'I don't give a monkeys. That'll settle their hash for now...it'll give them something new to think about!'

We didn't see either of the Grimms after that. They appeared to be keeping out of our way, and who could blame them? I did keep chipping away at Gordon, asking him to put the gate back on, but he wasn't having any of it. The lane was actually a public footpath: one that was very little used, admittedly, but without the gate at the top of the road I could foresee people starting to use it, as there was now somewhere for them to park their vehicles. Gordon did eventually admit that he'd been a bit hasty, and finally gave way to my nagging, by saying that he'd get it put back soon. I saw him up near where he'd thrown the gate, but nothing much happened.

An explanation came a few days later, when Gordon said, 'That gate is bloody heavy! I can't imagine how I managed to carry it that far, let alone throw it over the hedge. I can't actually lift it by myself, I'm going to have to ask one of the lads down the pub to give me a hand!'

We were then forced to drag ourselves, kicking and screaming, out to the local pub that evening! We met quite a few old friends, including Clive and Jayne, who were always extremely good company. Clive regaled us with tales of his latest exploit. After a heavy session at the weekend, he had slumped into bed, waking half way through the night for a call of nature. He didn't put the light on and, in his befuddled state, thought he was still in his previous house. They had only moved a few weeks before and he'd grown accustomed to feeling his way to the bathroom in the dark. Obviously, the new house wasn't laid out in exactly the same way, so instead of arriving in the bathroom, he somehow climbed out of a window, ending up on the roof of the extension built over the kitchen. Fortunately, the cold air revived him and he was able to regain his bed without mishap.

Just before we left the pub that evening, another mate of Gordon's hailed us. 'Hey Gordon, can I have a quiet word? Did you know that Dave Hill was in here earlier, asking questions about you?'

'Dave Hill? Don't think I know the name... are you sure it was me he was after?' queried Gordon.

'I think so...yes...I'm sure he said "Gordon from the barn". Be warned, he's a big bloke and very bad news,' cautioned Alan.

'Oh, well, I'm sure, if it's that important, he'll catch up with me sooner or later,' shrugged Gordon, dismissing the incident.

As we stopped at the top of the lane to open the gate, we saw the Grimms' lights being flashed on and off. We were being given the flashing treatment! The very bright lights just over their porch, and also the light at the side of the house, were both being switched on and off - like a signal. I don't somehow think that the message read, "welcome home"! Our other neighbours had long been treated to this spectacle, when they returned home after dark, but this was the first time for us. It looked as if we were well and truly in the doghouse this time.

After a very pleasant evening spent in the pub, Clive agreed to come and help Gordon retrieve the gate. Clive was a Martin Clunes look-a-like, who was never seen without his navy-blue shooting waistcoat. He and his waistcoat arrived in the yard early next morning. After a reviving cup of coffee, they set off together to weigh up the gate situation. I saw them poking around in the long grass and nettles, then Clive and Gordon heaved the gate up off the ground, staggering back across the road with it. They then struggled to locate the gate on to its hinges. While I was giving them another well-earned cup of coffee, Clive asked who owned the vehicle parked by the Grimms' yard gate.

'Don't think I've seen that one before, why?' queried Gordon. It was a distinctive silver 4x4 pick-up truck.

'Well...I've only ever seen one of those locally...it's owned by Dave Hill,' mused Clive.

'Funny you should say that, Alan said someone of that name was asking questions about me. His name doesn't mean anything to me...should it?' queried Gordon.

'He's a local...er...you could call him an enforcer, to put it politely. Into anything really. Even known to start shooting before he starts talking. A thoroughly nasty piece of work. I would say he's a very bad man to get on the wrong side of. He was almost brought up by the Luscombes, as his own parents didn't give him much time when he was a child. I should think he sort of thinks of James and George as his surrogate uncles.' Clive already knew the saga of the gate and after a bit of thought, he said, 'I used to go to school with Dave and knew him well, once upon a time. Think I might just give him a ring, to see what's up.'

'Do you think it's really necessary?'

'Yes...yes, I do as it happens. I have a nasty feeling about this. I wouldn't like to see you hurt, when I can perhaps do something about it. Trust me, a phone call wont go amiss.'

Later that day, Clive rang to speak to Gordon and, after a very puzzling one-sided conversation, gave me the gist of what Clive had reported.

'Clive says that Dave Hill was asked to come to see the Grimms, because I'd threatened them with violence. Lying little devils. Apparently the Grimms asked him to make sure I had a good seeing-to on a dark night but, now that Clive's explained matters to Dave and told him my side of the story, he's thinking about it.'

'You're having me on, aren't you?' I asked alarmed.

'No, I'm deadly serious. Clive reckons he's the local hitman,' he said shaking his head in disbelief.

So...George's words about knowing a man who'd bury Gordon six feet under, were coming back to haunt us.

'What's going to happen now?' I demanded.

'Not a lot, I should think, now Clive's put him right as to what really occurred.'

'I don't like it. I think we should ring the police and tell them.' I felt very edgy.

'No.' he replied emphatically. 'Stop panicking. Nothing's going to happen now.'

'But you said he was thinking about it. What if he decides to—'

'Drop it I said.' Gordon snapped. 'Just forget about it. It'll all come out in the wash.'

He didn't seem in the least perturbed; as if it was an everyday occurrence, being told you were going to be beaten up. I was worried sick and made sure that Gordon never left the house without me, especially at night. I'm not sure exactly how I thought I was going to forestall any attacker, but the thought was there. My mettle was never tested, as nothing ever did happen. This was more than probably thanks to Clive's intervention, but I never really forgave the Grimms their perfidy.

CHAPTER TWELVE
BIRTH AND A DEATH

The extreme ill will that had been generated by the gate incident, rattled on for weeks. The Grimms would ignore us whenever they possibly could and, if they should have the misfortune to encounter us in the lane, they would almost do a vanishing act, hiding behind whatever was available. We began to think they really were characters from a fairy story and that they were a figment of our imagination. It made us laugh, on more than one occasion, when we spotted them hiding behind a tree or peeping over a hedge to see if we'd gone. They seemed to think, almost like children, that if they couldn't see us, then we couldn't see them!

They didn't even appear to go into town so often these days: it was quite a rare sight to see the car trundling up the drive. I hadn't caught sight of Verity for literally weeks. If they didn't go into town, she could usually be seen peering out of the grimy window on the ground floor. We came to the conclusion that she must have had a chair set beside the window, in order to watch the world go by.

The grocer's van started to make a call once a week: it looked like they were, once again, doing most of their food

133

shopping with him. They had used the mobile grocer exclusively for their shopping, when we first came to live in the caravan. As they had crashed their car at the time, and had no other transport to get into town, they had no alternative, but rarely seemed to have used him since getting their wheels back. It was a mystery why, all of a sudden, they had stopped their visits to town and were staying so close to home. It was a topic of endless discussions between Gordon and I. The Grimms' predictable behaviour was an established fact but, when they acted out of turn, as now, then it usually meant trouble of some kind. And usually for us.

We were in the garden, just starting to lay some slabs onto the newly formed terrace, when George came into the yard. You could have knocked us both down with a feather! After craning his head around, looking to see where we were, he swaggered nonchalantly up to us.

'Us needs a strong fella-me-lad like you for a fu minutes,' said George jovially, pushing back his greasy hat further towards the back of his head.

Well, what a cheek! He was acting as if nothing had happened between us, almost as if the feud that had gone on for weeks had never happened.

'I'm busy, what do you want?' was Gordon's snappy, unequivocal reply.

'That little heifer of our'n be trying to calf. Old Jenkins the vet's 'ere now but, me and James, us is gettin' on a bit and we aint as strong as us once was. Old Jenkins is an old dodderer too, come to that, and aint as fit as he used to be.' George cackled.

We'd heard a lot of bellowing going on and, as the young heifer looked as if she was getting near her time had assumed, correctly at it turned out, that the noise was her efforts to give birth. I began to feel very sorry for her: the noise had gone on intermittently for a few hours now. She had only been a calf when we'd first moved to the barn and was hardly more than that now. The old bull had long gone and the only male of the species now on the farm was her own brother, so it seemed that a little interbreeding went on in the Grimms' herd.

'You'll have to wait if you want my help. I've got a load of concrete going off,' Gordon said unsympathetically, not giving an inch. George's face was picture of despair.

'I'll come, if you think I can be of any help,' I piped up. Hoping-against-hope that the situation wouldn't warrant it.

'We'em likely to lose her if we don't get someone else pullin' on them there ropes with us,' George warned.

'I'd better come, then, and see what I can do. I'm pretty strong, so we may be able to manage between us,' I replied with more confidence than I felt.

We set off together towards the shippon. George climbed through the five-barred gate into the yard.

'Wait for me George!' I wailed.

Standing in the yard, in front of the shippon, was the young bull and I was not in a hurry to reacquaint myself with him, if I could help it - even if he was the anxious father! The bellowing of his sister had greatly added to his agitation and he looked very wild-eyed to me, tossing his head and snorting through his flared nostrils. George shook his trusty stick at the bull, which made him back off slightly; enough, at least, for me to run across the yard and into the broken down doorway of the shippon.

It took a few seconds for my eyes to adjust, from the bright daylight, to the deep gloom inside. The scene before me could have belonged to a century or more ago. It was very dark, with only the two front doorways giving very little light. The heifer was standing in one of the stalls that would have been constructed from timber, many years ago. They were now broken and wormy, darkly coated with more than a century's build-up of muck. There, standing at the rear of the heifer, was James, with his shirtsleeves rolled up past the elbows, hanging on, for all he was worth, to a blue piece of nylon rope. One end of the nylon rope was looped round and round his ham like hands. The other end disappeared inside the rear of the heifer.

A stranger approached from the head-end with his hand outstretched. 'John Jenkins,' he declared. 'I'm the vet. You must be Mr Luscombe's neighbour. Thanks for coming to help. We're in dire need of another pair of hands here, as you can probably see.' He shook my hand. It was only then that I noticed, that both he and James were plastered liberally from head to foot in brown splodges.

George entered the shippon behind me, picking up the end of a trailing orange rope as he went past.

'I'm just going to attach another line to the other hoof, and then you can all start to take the strain, please,' said the vet.

135

I hadn't managed to utter a word by this time and could only gulp the lump in my throat further down into my chest. This was all unknown territory to me. I always had to look away when James Herriot, in the vet series on TV, was doing something similar to that going on here. Mr Jenkins ferreted away inside the heifer for a few minutes and then handed me the other end of the rope.

'OK everyone. On the count of three, all heave together please...one...two...three!'

Just as I was starting to take the strain and brace myself for the effort, the heifer sprayed all of us with a fountain of diarrhoea. *So that's what the brown splodges were*!

'Come on now, put your backs into it,' admonished the vet. We girded our loins and tried again.

The humans strained in virtual silence, only grunting occasionally with the exertion, while the heifer bellowed loudly, before the vet said, 'OK, let go for a minute, take a breather and let me see what's going on.' His arm disappeared again. Whilst trying not to look as he was doing this, I heard a car door slam. I was fairly near the door so went to see who had arrived: it was Belle's car. She was just unloading some more sheep out of the back seat, and shooing them through the gate into the field, to join the rest of her flock. 'I'll be back in a minute,' I said, 'I've just spotted a friend: she'll more than likely have time to come and help.'

I scrambled back through the gate, keeping a wary eye on the bull, which was standing with its head over the other doorway, watching the horrific events unfolding within.

'Belle...Belle,' I shouted, running as quickly as I could down the lane to where her car was parked. 'The Luscombes are having problems with a calving. The vet's here but they could do with some help, if you could spare the time. We seem to have been pulling for ages but nothing is shifting it,' I panted.

'Of course. I'll come now,' she readily agreed.

The vet greeted her with, 'Ahhh... good, just what we need, another pair of hands! We've got a very big calf here and she's only a small heifer.'

She *was* only a small heifer. A very young one too; straining to produce her brother's calf.

'Perhaps you'd like to hang onto this, my dear, and pull when I say,' requested John Jenkins.

There were five of us now, all straining on the word "pull", and resting in between. My legs and arms had now started to wobble with the strain when we were on a "rest". Unfortunately, our efforts didn't seem to be getting us very far. By now the poor heifer kept going down onto her knees and seem to be all but done in. The vet was starting to look rather anxious. Not having been present at a calving before, I was even more worried.

'I think she's getting a bit weak. She's gone a bit too quiet for my liking,' John Jenkins said. Turning to me, he continued, 'Do you think you could persuade that husband of yours to join us? Otherwise, I'm very much afraid we're not going to make it.'

'I'll go and ask him. I'll be as quick as I can,' I replied.

I raced back to the barn as quickly as my jelly-like legs would carry me. Dashing up the steps two at a time, bursting out onto the terrace, where Gordon was beavering away laying slabs.

'Gordon, please, you've just got to come and help...please...you've got to leave that...the poor animal is going to die in agony if you don't come,' I pleaded with him, more than a little upset, and rattled by the dramatic events unfolding in the barn. The tears were streaming unchecked down my cheeks.

'Calm down, calm down, woman. What do you need me for?' asked the voice of reason.

'We need another pair of hands. Even Belle is there helping but it's just not enough. We've been trying for ages. We need someone stronger: the bloody Grimms are worse than useless. It won't take a few minutes...please...please come,' I begged.

'Oh all right. I'm coming. Anything for a quiet life. Let's get it over with.'

He strode on ahead, with me trotting behind him as fast as I could go.

He shook hands and introduced himself to the vet; nodding hello to Belle, only giving a dark glance towards the brothers Grimm, who were still hanging onto their end of the ropes, grinning inanely.

'What do you want me to do here then?' he queried.

'Good man. Just catch hold of this, and when I say...start to pull. On the count...one two, three PULL! the vet bellowed.

I had only just taken the strain on my rope, when the heifer gave an almighty bellow: that extra pair of hands had done the trick. The calf shot out, feet first and landed, with a distinct thud, on the dirt-encrusted concrete behind its mother. The poor heifer chose this moment to go berserk. Her head had been tied, with a rope to a ring in the wall, to keep her upright, during the birthing; but she managed, with one mighty heave and jerk of her strong neck muscles, to pull the ring out of the wall. She then leapt into the air and, by an absolute miracle, just landed with her hooves on either side of the still body of her enormous calf. She looked absolutely demented, her eyes bulging, thrashing her head from side to side and bellowing loudly. She took in the sight of four men and two women barring her way to the light; making the decision to jump the timber partition of the stall. Her two forelegs immediately gained the top and, by scrabbling wildly with her hind legs, she managed to straddle her belly on top of the timber partition. After savagely bucking and writhing, she managed to clear the partition, gaining access to the next stall. The door at this end of the barn had only its top-half open. She stood weighing up the scene for a fraction of a second, before deciding that this was both too narrow and high to clear. She then turned towards the other doorway, where Belle and I were standing utterly rooted to the spot. I grabbed hold of Belle's arm, just managing to drag her out of the way, as the heifer thundered past. She gained the yard, sailing over the chest-high rails like a racehorse and shot off up the lane, as if the hounds of hell were at her heels.

In the furore, no one had taken any notice of the calf but, now that the heifer had made her exit, we all turned towards where it was lying. It had started to raise it's head, so was obviously alive, and after some attention from the vet, seemed none the worse for it's brutal birthing.

'It's a fine bull calf,' announced John Jenkins, 'It's a monster, no wonder we were having trouble. Well done everyone and thank you.'

Gordon, Belle and I said our goodbyes and left.

'Thanks for your help Belle, hope you haven't been held up too long,' I said.

'No, no problem. I'm off to get a shower, I think, I'm a bit messy!' she said, laughing ruefully.

'Mmm, me too, I definitely smell more than a bit farmyardish! I could do with a good scrub and a change of clothes. See you later then.' I replied.

Gordon went back to work for the half-an-hour left before it was time to finish for lunch. I went back to the caravan, discarding every stitch I was wearing onto a heap on the floor, outside the door. I then took a long hot shower, scrubbing off all the accumulated muck and gore from the birthing. Before his lunch, Gordon also went through the same routine.

After lunch, I felt as if I deserved an early day, so took the dogs for a long stroll through the fields. My leg and arm muscles still ached, but I felt euphoric. George was sitting on the lowest rung of the gate that straddled the entrance to the yard, when I walked past on my way home. There was no sign of the bull; all seemed calm and tranquil, the air drowsy with pollen and sunshine.

'Hello maid! I'm just sittin' 'ere enjoyin' a bit of a break, after that 'ard mornin',' George said.

'Yes, I know what you mean. I feel absolutely exhausted, but it's good to be alive today.'

He was still dressed in the clothes he had on that morning, complete with dung and bloodstains. The exposed areas of his lower arms and hands were also liberally smeared with this foul concoction, as was the whole of his face and neck. I horrified myself imagining him eating his lunch, with his hands unwashed, after the morning's work he had put in. He always had a pungent smell, but today the odour was much more in evidence - or was it just my fancy?

'Come and see the little un. He'em doin' right well,' he invited.

We both stood in the doorway, admiring the scene. The cow was standing, with the calf suckling for all he was worth, by her side. He was too large to fit underneath! I could see she was going to be hard pressed to satisfy that monster's appetite.

'Did she injure herself when she jumped the barrier and the rails?' I asked.

'Not too badly. The vet 'ad a look at 'er, when we managed to catch up with 'er. She was up the top, by them bungalows. We should enter 'er into a race, I reckon! She'em got a fu cuts to 'er legs, but they'll 'eal. She were lucky.'

'Yes she was. I thought she was going to mow Belle down. She was lucky too!'

'Aah, you moved pretty smartish, I have to say. Good job!' he chuckled at the memory. He was still picking the remains of his lunch from his teeth, with a piece of hay I'd seen him pick up from the yard floor.

The calf had finished its suckling by now; lying peacefully in the straw next to its mother. What a lovely sight they made.

'I could do with a drop o' some'ut to celebrate with...I don't know about you,' said George.

'I've got work to do, thanks all the same. You'll have me fit for nothing, if you start plying me with drink, at this time of day!'

'Go on, one won't 'urt yeh, let me give our James a shout for that there whisky bottle,' he cajoled.

'Oh.... go on then,' I gave in - very easily lead. These increasingly rare good relations with our neighbours needed a little nurturing after all, I reasoned.

After he'd shouted James a couple of times and received no answer, he said, 'Wait 'ere a minute and I'll go see what us 'ave got by way of a drink.'

He left me sitting on the bottom rung of the gate, very contented to wait, in that sunny spot. When he returned he said, 'Them bottles is all empty. I think James must a bin at it already! 'Ave you got a drop o' someut at your place?' he asked hopefully.

'Well, I know we haven't got any whisky, but there might possibly be a drop of brandy left from Christmas,' I cautiously said. I had work to do really, and could well do without a drinking session breaking out.

'That ud 'it the spot proper, maid,' George said, smacking his lips with relish.

'Come on then,' I said, resigned to my fate, leading the way across the lane.

I went inside to fetch the brandy, leaving him on the doorstep. I felt a little inhospitable doing that, but didn't want him inside wearing his crusty Wellingtons, and in the state he was in, blood-and-guts-and-all. We sat on the step outside the kitchen door with the brandy bottle between us.

'I haven't forgotten to ask Clive about those pheasants you wanted but, as far as I know, he hasn't been shooting for a while. It's probably not the season for it,' I said.

'That *is* right. Maybe when the autumn comes he might bag a brace or two. We do love a nice roast pheasant.'

'Well you shall have a brace, just as soon as I can get hold of them. What have you been up to lately?' I asked, more for something to say, than a desire for an answer.

'Nothin' much. Not been anywhere 'cause of our Verity,' he said reflectively.

'What's the matter with Verity?'

'Dun rightly knows but she 'aint right and no mistake.'

'I'm sorry to hear that. I wondered why you hadn't been into town much. You should have said, I could have done you some shopping if you'd only asked.'

'We don't do too badly. We got most of what us wants, but you could pick us up some cider and whisky, when you goes into town next.'

Now how did I guess it was only drink they might be short of? 'Yes, I'll do that. Have you had the doctor in to see Verity?'

'No she wont 'ave 'im, she's proper cussed, but she aint gettin' no better, say what you likes. I thinks us ull 'ave to give 'im a ring soon.'

'Good idea. He'll probably be able to put her right in no time.'

'She's that afraid she'll 'ave to go into 'orspital she just wont 'ere of 'im comin.'

'She's likely worrying for nothing. It is always worse than you think. Try to persuade her to at least see the doctor. If she's so poorly, it really is the best thing to do.'

'Aaah, I will and no mistake. Could do with seein 'im meself anyway. Me eyes is awful, missis.'

'What's up with your eyes?' I asked, turning to gaze into his well creased, but still lively orbs.

'I can't see a thing now. You see that there fence up yonder?' He pointed to our perimeter post-and-rail fence, up beyond the stream.

'Yes?' I said cautiously.

'Well I can't make it out from 'ere now, not no way,' he said triumphantly.

'You probably just need some new glasses. Come to that I haven't seen you wearing glasses at all lately.'

'They don't 'elp much but I broke the last lot a fu weeks ago. I can't see to even read that there newspaper now.'

'You more than likely just need your eyes testing again. It's free when you get to sixty-five you know?'

'Oo says I'm sixty-five?' he asked with an impish grin on his face.

'Oh, go on with you,' I said. It was difficult to tell but he had to be more than sixty-five. Didn't he? I was beginning to doubt myself.

'How old are you then?' I asked.

'That's got to be tellin',' he said coyly. 'How old do you reckon?'

'Sixty-eight?' I said erring very much on the low side.

'More or less, more or less,' he said grinning. I think he was well pleased with this guesstimate of his age. He thought himself a proper old dog. 'Tell you what maid, why don't you get me a pair when you go to that there car boot sale next time?'

'You can't just wear someone's old specs, you need to get your eyes tested properly,' I said, aghast at the thought.

'They'd be good enough for me, they would. I wore Ted's old pair but then they broke when I sat on 'em.'

The level in the brandy bottle was well down by the time Gordon packed up work and joined us. By the time George could be persuaded to leave, the bottle was totally empty. I had finally managed to convince him that there was nothing more remotely alcoholic to drink in the house. He staggered on his way, pleased with the day's work. All was well in his little world. He had been at his most charming and affable: I couldn't help feeling it was a great pity he was such a Jekyll-and-Hyde character. Our relations with the whole family could have been so much better, if only he were always in this amiable mood.

The doctor did come to see Verity. I saw him getting out of his car with his bag of tricks in his hand. I didn't see either of the brothers, but next day an ambulance rolled up and took Verity away on a stretcher, it's lights flashing, the siren echoing long after it had departed. Things must have been more serious than they thought. When I did see James and George quite a few days later, they were on their way into town.

'How is Verity?' I asked.

'She'em in the 'orspital,' James replied.

'Have they been able to find out yet what's the matter with her?'

'They'm doin' all sorts o' tests right now. We'em just on our way to visit 'er.'

'Make sure you give her my love; tell her to get well soon.'

'I will. You could go and see 'er yourself if you wants to. She'd like that I'm sure.'

'Oh, do you really think so? Yes, I'll certainly do that if she'd like me to. Ask her today, to make sure she's up to a visit from me. She may not feel well enough, and I wouldn't like to intrude.'

I saw them returning, very late that afternoon. As George was staggering almost legless down the lane, I thought it only prudent to avoid them. I was sure to catch up with them later, when they had sobered up.

Next morning, while I was in the kitchen washing up, George knocked on the window. 'Good morning.' I said, very much surprised that he's called round. It was usually only in an emergency that he actually set foot on our property.

'Mornin' to yuh. Went and saw sister yesterday and she'd like you to visit today, that is if you can spare the time.' He was unshaven and grubby. His dun coloured greatcoat held together with orange binder twine and his wellies very much the worse for their mucky yard: he looked more than ever like a scarecrow.

'Oh, good. Yes I'll certainly go. What time is visiting hour?'

'They don't just 'ave an hour now, you can go there any old time you likes.'

'Right, I'll go later on today then, probably around four, so it won't interfere with your visit.'

'Proper job, missis. I'll tell 'er when we goes later on this mornin'.'

Gordon came with me to visit Verity. I wasn't exactly looking forward to the visit as I'd hardly ever spoken to her. I could envisage the conversation flagging slightly. At least Gordon was a good conversationalist and seemed to be able to talk to anyone about just about anything.

She wasn't alone when we eventually found the ward she was on. Vince was there, a lady of about sixty and two young women in their twenties or thirties.

'Sorry if we're intruding but George and James said that Verity would like us to visit her today,' I said as we approached the group surrounding the bed.

'Of course, of course, no problem,' said Vince. ' Come and meet the family. This is my wife Lily and my two daughters Jane and Julie.'

We shook hands all round and then had time to take in the fact that Verity was propped up with pillows, wearing a fluffy pink bed jacket; her head was as bald as a coot. She was as pale as a ghost and, unless we'd recognised Vince at her bedside, I'm sure we'd have passed on by. This was the first time I'd had time to observe her closely, or seen her without the usual make-up she wore. She was awake, but looked a bit drowsy, and I got the impression that she had not long woken up.

'Hello Verity, and how are you feeling today?' I said gently.

'Mmmh, not too badly thank ee,' she replied in quiet voice. 'Oo be you then?' I'd bent over the bed while I spoke to her and she now had hold of my hand.

'I'm Maggie and this is Gordon. You remember us from Christmas don't you? We came and had a drink with you on Christmas Day.' She was still looking puzzled and, very obviously, hadn't a clue who we were! 'We live next door, in the barn. James, your brother, told us you'd like us to visit you.'

'The ward sister's just told me she's had quite a lot of medication. More than likely she doesn't even really know it's us, so I shouldn't let it worry you,' explained Vince. 'She's hardly said two words to any of us, in the hour we've been here.'

'Have they any idea what's wrong with her?' I asked.

'Not as far as I know. They've been doing tests for this and that, but the ward sister didn't know if the doctor had come to any conclusions yet.'

Not wanting to intrude on the family gathering, we chatted for a few minutes more and then made our escape. We were glad to get home; back to our own little world. All the same, we felt desperately sorry for poor Verity. It must have been a tremendous shock to the system, for her especially, being in hospital. Presumably they would have had to clean her up, as when we saw her at Christmas she had looked more than a little on the grubby side. Any time we had seen her, she was always heavily made up and wearing her wig: taking those away must have been very upsetting for her.

We did go again, a few days, later but she seemed no better than the first time we saw her, although she at least recognised us this time. We chatted away, more to each other than to her, about the weather, how the animals were doing and what the "boys" had been up to in her absence. We did manage to make her laugh when Gordon told her the tale of how James had asked to buy our chickens from us, after the fox had taken the last of his. Our small flock belonged to him in the first place and, to be honest, we were only too glad to get rid of them. They had given us magnificent eggs in return for their corn but were starting to worry us as to what would happen to them, if and when, we ever managed to get away on holiday.

We told James he could help himself to the whole flock. He and George came with a hessian sack apiece and proceeded to try and catch our three hens, six chicks and Corky the cock. The chicks were no problem and were very quickly sacked but the hens and the cock fought like demented things. They were all, eventually, ignominiously stuffed headfirst into the two sacks, still squawking and struggling, which the brothers flung over their backs and carried away. Every bit of the Grimms' exposed skin was running with blood, from the many scratches inflicted by the unwilling fowl: they looked more than ever like body snatchers.

We were never to see Verity again. Only the day after our visit she slipped into a coma and never recovered consciousness. It was a sad time for everyone. Her funeral was held at the local parish church, where Ted was buried and, like his funeral, had a good turn out. All the local farmers were represented by at least one member of the family, dressed very soberly in blacks and navy-blues. Weddings and funerals were always a good time for the local farming community to catch up with their neighbours' news. It was a beautiful service, held in a lovely old church, followed by the usual knees up at the Travellers Rest.

It seemed like the end of a chapter somehow. We knew that the only condition the Luscombes were allowed to keep the farmhouse was that Verity still lived there. If she died, or had to go into a nursing home, then they would have to leave. No one knew, of course, how or when this would happen but everyone had his or her own opinion to voice on the subject. Speculation in our small community was rife, the rumours buzzing along the telephone wires furiously.

CHAPTER THIRTEEN
AN EXTRAORDINARY PROMENADE
WITH THE SHEEP

The brothers continued in their new congenial mode and proved themselves to be affable neighbours, for once in their lives, at least as far as we were concerned. They really were mellowing. There were only the two of them left now and, although speculation was rife about their fate, life went on smoothly and placidly. There was no outward sign from the Grimms that they could be evicted at any moment.

One extremely hot afternoon, I saw their car returning from a trip into town. Normally, when they came home, George would be the gate man, opening and closing them while James, who was always the driver, drove down the lane. He would stop in front of the barrier constructed of wire mesh and old plastic bread trays, which was their Heath-Robinson style security at the entrance to the carport, waiting for George to come and dismantle it. That day, the car remained in the narrow lane, just outside the farmhouse gate, blocking it for any other car user. After hearing the blaring of horns, Gordon went to investigate, expecting to referee an argie-bargie going on between one of the people living in the bungalows and the Grimms. The white Ford car was empty, and not a sight nor sound from either of the Grimms. Very strange. Gordon helped Alex push the

Grimms' car out of the way, allowing enough room for his own vehicle to squeeze past, and then went to find out what had happened to them, just in case they were ill. When Gordon returned he said, 'The Grimms are having a funeral service.'

'A what?' said I startled.

'Well, apparently, they left the two Westies in the car when they went for a drink and, when they got back to the car, one of them was dead and I have to say the other doesn't look much better. I've told them to wrap it in a wet blanket for now and see if he recovers.'

'What a bloody witless thing to do! Don't they realise that a car gets like an oven if you leave it in the sun?' I felt sick and very angry, thinking of how the poor dogs must have suffered. It had been exceptionally hot that day.

'I suppose they've never had to leave the dogs on their own in the car before. Verity was always there with them, of course. You'd best not say anything to them; they're very upset about it. I'm sure they'll be a little bit more careful next time.'

They were both very fond of their little scruffy white Westies and, extraordinarily for farmers, treated them like lap dogs. The untimely death of one of them, must have hit very hard.

While the funeral had been going on in the Grimms' orchard, where they buried that poor little Westie, I had been studying our own orchard. As the fruit was just about ready for picking, I had been trying to get someone interested in taking the crop off our hands. Last year the harvest had been quite small; even so there had been little time, in any case, while the main building work had been going on, to spare for picking the fruit. What I had managed to salvage, had been used to make apple pies that I had sold to the Travellers Rest, with just a few going into our own freezer.

There were about fifty apple trees in our orchard; some more productive than others. There appeared to be an immense variety of fruit, from small sharp apples to large fairly soft textured ones. It was a real pleasure to stroll with the dogs through the orchard first thing in the morning, take an apple of my choice, and munch it on our walk. I couldn't identify any one type particularly but Ted had already told me some time ago, that a lot of them were rare and old varieties. My particular favourite was a small hard red apple with pinkish-white juicy flesh, which usually ended up running down my arms. Totally delicious.

I made many telephone calls before I finally found a local cider maker who, although couldn't take our apples himself, gave me the contact telephone number for a large cider maker. They thought that the larger company would probably be interested in taking our apples. Contact was made and I was very happy to find that they would take all the apples we could send up to their factory, just the other side of Exeter.

I started the picking early one morning while the dew was still on the grass. It looked a bit of a daunting task on my own. I started on the smallest tree; soon filling a sack with the ripe fruit, which I left propped up against the trunk of the tree. The Grimms had already given me instructions on how to pick and provided me with plenty of old scoured-out feed sacks. Their method consisted of beating the tree with a stick to dislodge the apples, and then collecting them from the ground. I knew this was the correct method from something I had read, but I didn't like the thought of the apples being so bruised, especially when they would have to wait in plastic sacks until I'd finished picking the whole crop, before being transported to the factory. So, whenever possible, I did pick them by hand and only used the stick method for the trees with the highest branches.

At the end of each day, Gordon would take the brimming sacks into the Atcost building, to keep them out of the way of any wildlife that had a fancy for an apple or two. Picking our crop was an enjoyable but tedious task. It was such a delight to be working out in the sunshine, however, the constant bending, when the stick method was in progress, eventually became backbreaking. The earwigs were something I learned very quickly to avoid. The first time I had a family of them in my hair, there were more than a few shrieks coming from the direction of the orchard.

The job finally took me five days to accomplish and it then took Gordon two of our trailer loads to get them all to their destination. A final weigh-in of almost one and a half tons was our tally, for which we received a grand total of seventy-one pounds. It was a good job there were no wages to pay out of this largesse! It was better than seeing them rotting on the ground, at least.

We were left with a few windfalls and some nice apples that I had especially kept for our own use. Some of these I made into apple pies for the pub and a small amount to be put in the freezer. The windfalls I crushed with a tiny cider press that Gordon had bought me for my birthday. The demi-johns of the resultant apple juice were stacked inside the utility storeroom and left to bubble away in peace.

The weather had been dry for weeks; the grass in the fields had started to get a little yellow and sparse thanks to the lack of water. Belle had decided to take her sheep away from our land, to give the grass a rest, and had taken them, two-by-two, on the back seat of her car, down to the other pasture she rented by the Travellers Rest pub. The Luscombes' land had also gone very dry. They grazed their sheep and cattle in the same field, which meant that at this time of year they had also been supplementing their feed with hay and bread.

To give their field a bit of a rest, they would occasionally bring the bull, the cow and her newly born calf, into the enclosed yard, feeding them entirely with vegetable peelings and bread. The hay they'd had originally stored in our Atcost building, had now all been used up, and no more was purchased to replace it. We took to saving all our kitchen waste, at least anything that the dogs didn't eat, and would place it just inside their yard. The cattle got to know when we were coming with a little extra tucker and would thunder out of the barn, all vying for the meagre scraps we would have. The peelings from my vegetables got thicker and thicker, just to give them a little extra. If I had the time I would hand feed the calf and then give the remains to the cow. Left to their own devices, the bull would shoulder them both aside to get all the choice bits for himself.

George turned up on the doorstep one morning. I wasn't aware that he was standing just outside the door, as he hadn't knocked and I hadn't heard him mooching around. When I went outside to hang out the washing, I discovered him standing there, gently tapping his stick on the ground, and looking round him in a very nonchalant manner. 'Ohhh, George, you did give me a fright,' I said jumping a mile. 'What can I do for you?'

'I be wonderin' if that boyee of yourn could give us a 'and with them there sheep tomorrer mornin'.'

'But it's Sunday tomorrow?' I said. Sunday was a strange day for them to be doing anything serious as it was usually a day they did very little, except garden. Sunday was supposed to be a day of rest, or so they had informed me, having told us off many times in the past for working on a Sunday.

'Yes, it be nice and quiet on Sunday. Sunday is a good day.' George reiterated.

'I'll go and ask him, hang on a minute.' More than a little intrigued, I went willingly enough to find Gordon.

'George is on the doorstep and wants to know if you can give him a hand to do something with the sheep tomorrow.'

'Do what, exactly?'

'I don't rightly know but, apparently, Sunday is a good day to do it!'

Gordon followed me back into the kitchen. 'What is it you want help with, George?'

'Them there sheep aint got no more grass and we aint got no more 'ay. I arst that old Mrs Nick Nock and she says we can put 'em on 'er pasture for a fu weeks until she wants it back ag'in for 'er 'osses.'

'How are you going to get them all down there? Have you got a lorry coming?' Gordon queried.

'Nah, there's no need of that!' George said scathingly. 'We can do the same as we allis do - drive 'em down there.'

'How do you mean?' Gordon said, puzzled.

George sighed and rolled his eyes at the sheer stupidity staring him in the face. 'That's what we wants you for 'o course! You drives in front of 'em with that there Land Rover and me and James will follow behind, like, with the dogs. Proper job and no mistake!'

'What about any traffic that might be on the road?' Gordon asked, somewhat taken aback.

'If we goes early enough, there won't be none!' reasoned George.

'All right, if you say so. What time do you want to start?'

'Can you be out of bed by seven?' George's impish grin widened on his grizzled old face. He looked as if he hadn't shaved for a few days and his neck was as thin and sinewy as a turkey's. Mrs Nick Nock had told us that he'd fought off cancer twice already, although he'd never mentioned a word of this himself. He must be as hard as nails, despite his age. Actually, he didn't look too healthy, when I came to look closely, but his indomitable spirit appeared to overcome all adversity.

'Cheeky! Yes, of course I'll be out of bed by that time! I'll meet you up by the top gate then.'

'Tis 'ansome me beauty,' George said, waving his stick in farewell.

As Gordon drove out of our yard next morning, the Grimms were already up at the top gate. They both had their greatcoats on, complete with orange binder twine belts and floppy-brimmed hats. They had a tall staff each and there were three dogs to accompany them. This looked as if it had the makings of a fine adventure, so I decided, there and then, to go too and quickly got into my wellies and waterproof jacket. The last few days had been very warm but, at that time in the morning, it was still a bit nippy and it looked as if it could rain, given half a chance.

The top gate was opened and Gordon drove the Land Rover through onto the main road. All, so far, was quiet traffic wise. The Grimms whistled up their dogs and very quickly rounded up the forty odd heterogeneous sheep from their pasture. There were black ones, white ones, black and white ones, raggedy ones with lumps of fleece hanging off, almost bald ones, big ones and little ones. Where on earth did such a bunch come from? They were like the dogs - the ugliest lot of sheep I'd ever seen.

They seemed to know what was expected of them; very quickly trotting after the crawling Land Rover, with its hazard lights flashing prettily. George crossed to the far side of the main road, to stop them going in the opposite direction, while James herded the last of them through the gate. I closed the gate and followed on behind, holding both arms out to stop any escapees that might decide to turn tail. A bit of chivvying up was sometimes needed, when small bunches of them occasionally decided that the luscious roadside grasses and wild flowers were far too good to pass on by; standing munching contentedly for as long as someone would let them. I found myself a piece of stick long enough to tap them on the rump to get them started again down the road.

The flock of sheep straddled the whole width of the carriageway. Although the vast majority of the road we were to traverse had long straight stretches, there were a couple of sharp bends that had my heart in my mouth. It was a main road after all and not the sort of terrain a person might expect to find a flock of sheep. I was a little concerned in case a car should be travelling so fast it was unable to stop, when faced with a barrier of sheep across the width of the carriageway. I heaved a sigh of relief after each bend had been negotiated.

I quickly got into the swing of this novel sheep drive and the jargon that went along with it. "Hup there me beauties" was my personal favourite. Just a tap with my stick on the road behind the sheep would soon do the trick; moving them along nicely. I felt quite an old hand at this animal-moving lark already.

Gordon had by now reached the lane that led up to the Nick Nock's place. I was wondering how he was going to manage if the sheep decided to go straight on, however, they were wiser than that, and followed obediently behind the Land Rover, like lambs searching for their mothers. The line of cars, patiently waiting to pass, might also have been a bit of a deterrent to any stragglers that might have decided to head for Exeter, instead of their nice new pasture. By the time I reached the Nick Nock's field with the last of the flock, the leaders were already gorging themselves on the lush new grass.

This Sunday stroll was to be repeated in reverse three weeks later. The sheep had decimated the field and stripped it almost totally bare by the time we came to collect them. They gave the impression that they knew the way home and the return journey was accomplished without any hitches. Gordon and I were now old hands at the game, knowing what was expected of us, too. It amused me when one of the car drivers we held up, shouted in a friendly fashion to me, perhaps thinking that they were my flock, 'Nice day for it missis.' The sheep seemed glad to be back, but took a bit of persuading that they must go into their old field. The grass was still very sparse there and they found the grassy knoll at the top of the lane much more to their liking.

That evening, in the Travellers Rest, Clive presented me with a brace of pheasants. Lovely plump birds that I knew would make the Grimms' mouths water. 'Thanks Clive, what do I owe you?' I enquired.

'Mmmh, let me think? What I would really like is some more of your delicious pasties,' he requested.

A few months previously we had all been having a conversation, gathered comfortably around the bar in the Travellers Rest, about Cornish pasties. Everyone seemed to love them but all had their own opinion as to which manufacturer made the best ones and exactly which

ingredients should go into them to make them authentically
Cornish. Then we progressed to Devon pasties. My
contribution to this conversation went along the lines that a
commercially baked pasty could not possibly taste as nice as
a home made one. This was where I was asked to put my
pasties where my mouth was! Clive challenged me to come
up with my best effort. Rising to the gauntlet thrown down, I
think I surpassed myself. The pasties were made with the
only the very finest ingredients: golden short crust pastry,
with just a hint of mustard and cheese, filled with tender
chunks of juicy steak. The end result was a triumph. Clive
had not let me forget just how good they were and, whenever
we met, would ask when he was going to get another supply.

In answer to his request I said, 'Then that's what you
shall get my boy!' Very glad that his wish was such an easy
one to fulfil.

The bag, with the brace of pheasants in, was put on the
floor, under our feet and, I am afraid, when we left the pub,
that's where they stayed.

It wasn't until three days later that we actually
remembered the pheasants.

'Good grief, the pheasants!' I suddenly said.

'Oh, no!' I'll have to go down to the Travellers Rest right
away and get them,' said Gordon.

'They'll have been thrown away, surely? I should think
they'll be stinking to high heavens by now, if not.'

'Well...pheasants are supposed to be hung, aren't they?'
said Gordon with a huge grin on his face.

'Yes, but not in a plastic bag! Imagine what the cleaner
must have thought, seeing a carrier bag on the floor, then
opening it up to find a couple of smelly dead birds.' We were
both laughing by this time. 'They'll be fit for nothing but the
bin. I suppose we had planned on going down to the
Travellers anyway this evening, so we'd better ask what
happened to them. That's if we remember...'

When we enquired later that evening the barmaid, Lucy,
told us that she'd spotted them by the side of the bar a few
nights ago and had hung them up on the back of the kitchen
door. As I predicted, they certainly did hum! We left them
hanging in the Atcost that night – up high enough that any
wildlife wouldn't be tempted.

Gordon presented them to George the next day. George
peered into the bag with his rheumy myopic eyes, 'Proper

job, boyee. Them's two for the pot!' He seemed very well pleased with his brace of pheasants and hadn't seemed to notice any untoward smell.

A few days later he stopped me to tell me just how well they'd roasted and how much they'd enjoyed them. 'Them pheasants were bootiful,' he said masticating his teeth, with the memory still fresh.

'Oh, good. I'm glad you enjoyed them,' I replied.

'James roasted 'em with some tatties and parsnips. That there meat just fell off the bones. 'Ave you got any more?' he enquired hopefully.

'No, but if Clive goes shooting again and offers me any, I'll certainly give them to you.'

'Proper job maid.' He was well pleased with the thought of more pheasants to come.

In between the sheep walks, we were trying to create a garden. The barn itself was entirely finished, inside and out, and we were very satisfied with the way it looked. We had already put terracotta tubs, overflowing with trailing lobelia, petunias, and sweet smelling dianthus, outside the front door, which faced onto the lane, but wanted to create a courtyard garden at the back. The ground sloped considerably from the top end of the barn and we had decided to terrace the garden in three levels. There would then be a connecting flight of steps to reach the lowest level, which served the backdoor, and the entrance to the Atcost building. Gordon had already laid soft creamy-coloured slabs on the top terrace and had built two small retaining walls, in stone, where the terrace descended. On the ends of each wall, he had included square planters that had been filled with roses and other small bedding plants.

We now started working on the lowest level, outside the kitchen door. This area was approximately fifty feet by twenty feet. The flooring was the original yard concrete, which we had found too daunting a task to take up, consequently had left it as it was. The condition of it wasn't brilliant; it did have a few cracks here and there but the alternative was unthinkable. We had built planters at each corner of the Atcost, which would eventually be planted with Virginia creeper, to hide as much of the building as possible. We were, first of all, cladding the face of the Atcost barn with diagonally laid shiplap boarding, which we intended to stain a dark oak colour. Gordon was sawing a piece of pine ready to be tacked onto the ones already laid, when we became aware of an altercation going on somewhere in the lane. The noise of the quarrel had been steadily growing louder until it became impossible to ignore.

'What's going on now?' Gordon whispered, keeping his voice down in case we were heard and drawn into the quarrel.

'It sounds like the Grimms and another man but I can't tell who it is, can you?' I whispered back.

'No...no idea. Let's go and see what they're up to.'

We both crept quietly up the steps leading to the top terrace; peering around the end of the barn wall. The scene that met our eyes was fairly unbelievable. There was George, rolling on the ground in the dust of the lane with our neighbour, from the farthest bungalow, down on the ground with him. Both of them were punching and kicking ferociously at each other for all they were worth. It looked like a playground scene but instead of five-year-olds with short trousers, here was one very old man and another, not in the first flush of youth, indulging in fisticuffs. The fight didn't last more than a few minutes, both soon regained their feet, shaking themselves down and generally regaining their equilibrium. It didn't take long though before the verbal disagreement continued.

'You...you...you bleddy gypsy,' George gasped. 'I'll get even with yuh if it takes the last breath in me body.' He stood there squaring up, with both his fists raised.

'I've had just about enough of you old man. It's about time they evicted you and collected all those bad debts,' said Alex between gritted teeth, advancing again towards his adversary. A rivulet of blood had started to drip down his cheek.

George dodged quickly inside his own gateway, standing out of harms way; shaking his fist and hurling abuse at Alex, who was walking back to his own house, further up the lane. 'You just wait.... my time will come. I may be old but I've still got me strength. I'll see you six foot under yet, see if I don't.' *Hell, he's going to call the bad boy in again!*

We'd already been told that all the Luscombes were fighters, especially George, but hadn't thought he would still be up for it at his age. Well, well, there was certainly life in that old dog. What on earth would they get up to next?

It was a few days before we finally managed to hear both sides of the story. Each of the pugilists were at great pains to give us their own version of events, putting themselves forward in the best possible light, denying hotly that they started it. Definitely five-year-olds in long trousers!

Alex was really embarrassed about the incident. 'I shouldn't have let him rile me. I knew it would happen but I just *had* to go and talk to them. I couldn't let it rest.'

'What happened?' enquired Gordon, not letting on that we had witnessed the tail end of the incident.

'They were both shouting and cursing at Kathleen when she came home from work; wouldn't let her pass in the car. They actually sat on the bonnet, waving their sticks at the windscreen. It certainly upset her. I think she was scared they were going to bust the windscreen in. The gist of it was, that she always left the gate open. The usual endless rubbish. They'd obviously had a few to drink and got themselves all worked up about nothing. It wouldn't have bothered me; I'm too used to it. I don't know why it bothered Kathleen, come to think of it; she should have been used to it too. I think it was the threat of physical violence that did it for her,' explained Alex contritely raising his eyebrows. 'It scared her silly. She was shaking all over when I saw her.'

'I'd better have a word with them again, I suppose,' decided Gordon. 'At least get them to lay off the women.'

'Would you? I'd appreciate that. I don't give a damn what they say or do to me but I don't want Kathleen upset again. I had a mind to get the police involved but they're never much use and, in any case, I'm as much in the wrong now as they are, I suppose. Just look at these bruises!' He rolled his sleeves up to show us ugly purple bruises down his arms. 'Look what fighting at my ages does for you! I'm as stiff as a board. I can hardly move without wincing. I shouldn't have hit him really, a man of that age, but he riled me up so much I just lost it,' he explained.

'No, not such a good idea letting them get to you like that.' Gordon always had a cool-as-cucumber approach to the Grimms. Well...most of the time anyway! It used to madden them even more. They would be shouting and cursing, while Gordon would come forward with a reasoned argument, put forward in a nicely controlled voice. Always a difficult one to deal with, and wrong-foots the opposition. I should know - he does it to me too. 'Leave it to me, I'll have a talk with them,' said Gordon, reassuringly.

Gordon didn't have to seek out George. He came to see us, with his own version of events, later that day. 'That gypsy, look what he's done to me,' he said. He still looked to be covered in the dirt of the lane, his face and clothes smeared with grey dust. He had a black bruise developing nicely down one side of his face but, strangely enough, looked in much better shape than Alex had.

'What have you been up to now,' asked Gordon, deciding to let George dig his own pit.

'Bleddy gypsy hit me, didn't he,' replied George standing belligerently with his stick swishing backwards and forwards, chopping off the tops of the dandelions growing on the verge.

'Hit you? What for?'

'He's bleddy mad that's why. Stark staring mad! The bastard should be in Digby.' (Digby was the local hospital for the mentally ill.)

Obviously this conversation wasn't going anywhere without a little prompting. 'Come on! He wouldn't have just hit you without a reason, surely? You must have provoked him.'

'Started shouting at me...just like that. Told you – he's as mad as a bleddy March hare!' George was still swishing at the grass and weeds, not looking Gordon in the eye.

'Rumour has it that you and James were having a go at Kathleen, that's why he was angry. What've you got to say to that?'

'Well...' He grinned roguishly, 'I might just have had words with that one too!'

'Look, George, leave it out will you. You frightened her half to death. She was very shaken up when she got home and that's why Alex lost his rag. I can't say I blame him really. You should leave the women out of this, they get a bit upset at your rantings.'

'Can't help that, boyee. Them's just rubbish and needs putting out!' There was just no amount of talking would ever make any difference to the enmity that had always existed between the two families that occupied the bungalows and the Luscombes.

'Do me a favour and lay off them will you? It's me that gets it in the neck when you've been naughty boys. Alex was all fit to get the police involved, until I talked him down. Let's all try and live together a bit peacefully, shall we?' reasoned Gordon. 'You know it makes sense.'

'I aint promising nothin'. If he leaves us un alone, then maybees there wont be no trouble,' was the uncompromising reply. He left, stabbing his stick aggressively into the ground as he walked.

Events were soon to overshadow this long-standing feud between these old adversaries and, in the next few months, the Luscombes would have more on their minds than causing trouble for Kathleen.

CHAPTER FOURTEEN
A SHARP OBJECT LESSON

Gordon had been incarcerated, working on the barn, for so long that he had almost forgotten that there was a big wide world out there. We had had little time for leisure; there had been more important issues for us to contend with, but the job was now coming to an end and it was time to enjoy ourselves a little. We started to have a day off a week, not necessarily a Sunday as this did not invariably fit into the scheme of things, but just one whole day, free to do exactly as we liked, was a real luxury. We started to explore the surrounding countryside, taking the dogs with us for long walks, usually ending up in a snug pub somewhere for a good lunch and a few pints of real ale.

One of our favourite watering holes was the Nobody Inn at Doddiscombeleigh, an ancient old Inn that was famous for it's cheeses, whiskies and wines. Our favourite lunch was their sausage and mash with onion gravy. Three fat brown sausages were served on top of a nest of fluffy mash potatoes, together with a small jug of gravy, and a tiny pot of grainy mustard. It makes my mouth water just to think of it.

It was also very enjoyable to walk along the Exeter canal, ending up at the Opposite Locks pub; especially when the weather was fine enough to sit outside against a warm wall, enjoying a pint of real ale. My tipple was a half-pint of Merrydown cider, which always seemed to go down well, especially after a long hot walk. It was the sort of inn where anything could happen. There were always children diving or swimming in the canal, or playing on the many bits of timber apparatus erected especially for them. No one seemed to mind dogs roaming around, it was so free and easy, and often a group of musicians would tune up and start to play.

The very best walk of all though, for us, was the walk starting from Fingle Bridge over the river Dart. We would walk up river as far as we could manage, sometimes taking in a detour up to Castle Drogo along the way. We would then retrace our steps and end up for lunch in the Drew Arms at Drewsteignton. Absolutely fabulous. The first time we attempted this walk there was a party of eight including the two dogs. I was chatting away and looking around at the wonderful scenery, not taking too much notice of what the dogs were up to. There was no need to keep a strict eye on them generally; they were obedient and tended to stick with us, used to behaving well around farm animals. They must have been a little impatient this particular day and had run on ahead, with two of the men in the party. Suddenly there was a splash, and in the distance I could see Sharp in the water. This was accompanied by shouts, but I knew immediately there must be a problem, as Sharp absolutely hated water of any description. He was being swept rapidly downstream in the strong current, with all of us running behind trying to head him off. We were very fortunate that he didn't in the end need help, thank goodness, because none of us would have been in time to get anywhere near him, as the river was in full spate and the weir very close.

Only after he had hauled himself out, shaking water everywhere was I able to find out what had happened. Just at the river crossing, there was a metal bridge with openwork steps leading up. When one of the men had got to the foot of the steps, he tried to encourage the two dogs up. Both were extremely reluctant, but Tess eventually made her way up and over to the other side. Sharp, unfortunately had his own agenda, deciding immediately that he didn't like the look of the steps at all, shooting down the steep bank and leaping into the water. Sharp had spent the first ten years of his life confined to a farmyard so it was unlikely that he had encountered a river before. Especially a river as fast flowing as this one. It was a good job he was a strong dog and managed to swim, albeit almost sideways, to the bank. I think he learned his lesson after that because whenever we went on that particular walk he would wait at the bottom of the steps for someone to pick him up and carry him over the bridge.

We had an unexpected afternoon off when we went into Newton Abbot with the Land Rover and trailer, one Sunday afternoon to collect some materials from Trago Mills, a large do-it-yourself store. On the way there, we noticed a poster that told us a beer festival was being held at the Maltings. The Maltings was literally a large stone building where the barley was still malted for beer and had now been turned into a fine venue for a real ale festival. Gordon, being a connoisseur of a good jug of ale, said, 'How about going for a pint after we've picked up the gear?'

'Are you sure that's a good idea? You know what you're like when you get a pint inside you. It'll end up as two, for sure!' I said.

'Oh come on woman, I'm only talking about a quick pint for goodness sake! It's not the end of the world!'

'All right then, but don't forget you're driving,' I reminded him.

The materials only took half an hour to buy and load onto the trailer; we were also lucky to find a suitable parking place near to the Maltings, which looked very busy. Inside the venue there was a row of about thirty barrels, all with white cards attached to the tops giving the name of the beer stored inside and a short description of each one. Which one to choose? No problem in my case, as I didn't like beer that much, but it took Gordon a few minutes to make his choice from the superlative West Country ales on tap. It went down a real treat and it was good to see him enjoying himself. As predicted, it didn't end there and he had to try just one more. And then "just one for the road". And then "one for no reason".

By the time I could persuade him to leave, he was in no state to drive. Being stone-cold sober, it made sense for me to drive but, never having driven the Land Rover with the trailer attached, I was more than nervous as this was a twin-axle tipper trailer, as big as the Land Rover itself. This was going to be a challenge. Gordon sat beside me, smiling

161

benignly on the world, patting me on the arm every now and again. 'You're doing great,' he slurred. 'It's a piece of cake. Just remember it's only as big as the Land Rover and just forget all about it.'

Ha! The first roundabout I came to made me realise it couldn't be forgotten about! I careered onto it, and then off it, scaring myself silly. I would have to be more careful and slow down a little for the next one. By the time I got to the motorway, I was a nervous wreck. Gordon was, blessedly, asleep and only woke up, with a startled snort, as I stopped at the top of the drive to open the gate. At least we got home without any mishap.

This was a period when the work on the barn had come to an end and Gordon was looking around for something to occupy himself with. Henry, who had helped Dawson make, and fit our windows and doors, had his own carpentry workshop in the village and, like all good craftsmen, was inundated with work. Gordon had been asked by Henry to take on a job that he himself had no time to do. It was the refurbishment of a forty-two feet long cabin cruiser. Gordon went with Henry to see the boat, which had already been taken out of the water, and was lying in a field only a few miles away. They agreed between them that the boat would be towed from its present grassy berth to our Atcost building, until such time as the owner had time to discuss his requirements with Gordon.

A purpose-built trailer had already been made, by Clive, for the boat to be transported on, and to rest on when it was inside the barn. An enormous tractor, with tyres that towered over my head, towed it to the edge of our field via a neighbour's gateway from the main road. We thought that the bridge over the stream wouldn't take the combined weight of the boat and trailer, let alone the tractor, so had sought permission from our neighbour to go over his land. Part of our fence was removed to give the gigantic rig access to the field, where it made its stately way down to the yard. It was certainly a nerve-racking sight seeing it slide into the Atcost, inch by inch, but was halted at the last minute by a shout from one of the men. The height of the boat on the trailer was very critical and, as it turned out, too high. A section of the cladding was sawn away and at last the boat made its way under cover.

Gordon worked on the boat most days, mainly removing rotting timbers and other accumulated rubbish that lay in the boat. It was quite a few weeks before the owner put in an appearance. Unfortunately, his perception of what he wanted the completed boat to look like was just not practical and his pockets were not large enough to accommodate his ideas. After much soul-searching, he decided to scrap the boat and Gordon was asked to take off anything that was salvageable. The engines were definitely worth saving but, as they were seriously heavy, thought had to be given as to how they could be removed.

Gordon continued to work on the boat, but in a desultory sort of way, and when he had a phone call from Mike Madge asking if he wanted some work, he jumped at the chance. Mike was a Plant Hire Contractor who Gordon had worked for in the past, normally carpentry work, but also anything else that needed doing. Mike was currently working on a large demolition contract at Trusham Quarry, in the Teign Valley.

Gordon left early each morning, with a good lunch box beside him in the cab of the Land Rover, not return until quite late in the evening. He was as happy as a sand boy. And so was I for that matter! In the main we lived and worked together for twenty-four hours a day, without any animosity, but it was always good to have a few hours apart, every now and again.

Towards the end of the demolition job, Gordon had been talking about some of the articles that had been salvaged. He was particularly interested in a crane, for some reason. This was a gantry crane that was normally used for lifting engines out of excavators and trucks but Gordon thought it might be useful for lifting the two heavy engines out of the cabin cruiser. It was approximately eighteen feet high, twenty-one feet long and six feet wide, massively constructed of steel on iron wheels. 'What do you think? Should I buy it?' he mused.

'Don't ask me! What on earth do you want that for anyway?' I asked.

'It'd be useful for getting those engines out of the boat for a start,' he replied.

'Yes, maybe, but when it's served its purpose, what do you do with it?'

'Oh, I'm sure it'll be useful for something or other,' he muttered vaguely.

'Buy it then, if you want it. We've certainly got plenty of room to store it!'

'I might have a bit of a problem getting it here, that's the only thing. It's a big piece of kit. I'm going to have to have a serious think as to the logistics of transporting it,' he pondered.

In the end, after a few forays down to the Travellers Rest involving intensive conversations with Clive and Henry, a solution was found. Knowing that our own trailer would not be man enough for the job, Gordon loaned a trailer that was the largest legally allowed on the road. The crane was strapped down rigidly to the trailer but, looking at it, Gordon was not at all sure whether the load met the legal height requirements. He decided to take the back roads from the quarry, rather than the motorway, to avoid any unwanted attention - just in case. He told me later that when he looked in the rear view mirror, as he set off from the quarry, he could see immediately that the sheer height of the contraption was going to be a severe problem. Within five minutes of setting off, there were already branches of trees being ripped from their trunks, and he was on tenterhooks every time he saw an overhead cable, wondering if he would clear it, or bring it down.

He was driving very cautiously, taking each bend slowly, when in the rear view mirror he saw a police car - with all lights flashing and sirens going - coming up behind him. His first thought was - had he bought a power line down? He was not left to speculate for long, as the police car screamed past, on its way to a greater emergency. His relief was most definitely tangible.

He was like a kid with a new toy as soon as he got it back into the Atcost. He telephoned Clive, who had offered to give him a hand unloading it, and as soon as he drove into the yard Clive declared, 'I've got just the job for that.'

'What've you got in mind?' asked Gordon.

'It's just what we want for that paper mill in India I've been telling you about,' he replied.

'How on earth would you get it to India? asked Gordon in amazement.

'Mmmm, I'll have to make a few phone calls, but I don't think it will be too much of a problem.'

Clive made those phone calls. The gantry crane was stripped down, loaded into a shipping container, and eventually transported out to India by boat, for use in a paper mill that Clive's company was equipping with machinery. Gordon never did use it to lift out those massive boat engines. Instead the whole boat was sold by the owner, and collected by the same tractor that had brought it to us in the first place.

The Grimms had been quiet lately and we were being lulled into thinking that their new mellow mood was going to last. They must have had a gargantuan session in some hostelry or other before returning home, as it was plain from the first that George, on opening the gate, could hardly stand. He stood there swaying gently, hanging onto the end of the gate, while James drove the car through. When it came time to close it, he had a problem for a few minutes arranging his feet. Eventually he sorted himself out, making his unsteady way down the lane, behind the slowly moving car.

Gordon was half way up the drive, on his way to take the dogs for a walk, when he first caught sight of them. James wound down his window and I could hear that an almighty row was in progress. It didn't take George long to join in. The decibel level increased, making it easy for me to hear the majority of the words, mostly of the four-letter variety. This was something I didn't want to get involved in: making myself scarce in case they saw me and expected me to join in too.

When I caught up with Gordon, he was thoroughly disgusted. 'They are just about as legless as you can get. How James managed to drive back in that state, I just don't know,' he shook his head in disbelief.

'Do I really need to know what it was all about?' I queried.

'Hmm, no you definitely don't want to know. It was just the usual: nothing new. Why on earth do we have to put up with them?' was his parting shot.

Next day we saw a familiar car making its way down the drive. It looked as if the Luscombes had called the vet out. He was not a frequent visitor and I suspect, as with most farmers, they only called him out as a last resort when their own homemade cures had failed. *Wonder who was in the wars now?*

We were not left to speculate long. George coming through the yard gates shouted, 'Ow do boyee!'

After the row the day before, we could only marvel at his sheer effrontery. What the hell do you want?' demanded Gordon.

'Come and give me a 'and to push that there crush over to our yard,' he said, seemingly oblivious to the chilly reception his words had received.

'Why?'

'That there vet wants it so's he can take that young bull's balls off. He'em a bit on the frisky side and we could do with that there crush to make the job a mite easier.'

'Well you'll have to do without it then, wont you?' replied Gordon walking off. *Oh dear...*

George stood for a moment scratching his scalp under his hat. 'What's got into 'im?' he asked perplexed.

'You should be able to answer that for yourself!' I replied.

'Me? What 'ave I done now?' He shook his head in bewilderment.

'Don't you remember that tirade yesterday, when you got back from town? All that shouting and arguing?' I prompted.

''E don't want to take that to 'art now do 'e?'

'Nothing to do with me, George, but if I were you I'd try and keep a bit quieter when you get back from town in future.'

'Let's forget about all that nonsense. Come and give me a 'and to push that there crush,' he wheedled.

'Leave it where it is,' I warned, 'or there will be serious trouble. I'm not interfering in your rows. You'll just have to manage without it.'

George left the yard muttering and not best pleased. The next emissary was the vet himself. 'Sorry! There seems to be a bit of a mix up. Mr Luscombe has just told me that you wont let him use your cattle crush but, I would be very grateful, if you could see your way to letting me have it, just for a few minutes.'

'Yes, of course! There was a bit of a row here yesterday: I think Gordon was trying to tell George that he couldn't just walk in as if nothing had happened, just because he needed something.' I tried to explain, feeling more than a little foolish.

'Quite! Mr Luscombe's temper *is* a little unpredictable isn't it,' he said, the master of understatement.

The crush proved to be so heavy that I had to get Gordon to tow it over to the Grimms' yard where they had the young bull cornered. *Poor chap!*

Our second Christmas was looming up fast. I took immense pleasure in taking time out to enjoy and prepare for the festivities. This year it was the turn of Gordon's family to visit. There would be Gordon's eldest sister Anne with her partner Wal arriving from Canada; brother David, his wife Pat and son Alex; second sister Judith with partner Terry, sons Duncan and Brandon; and his youngest sister Jane, husband Ian, sons Andrew and Robin - all from Shrewsbury. In total, there would be fifteen this year for Christmas dinner.

The majority of the visitors arrived on Christmas Eve; consequently, this year there was no great build up to the day itself. I had been happily preparing for weeks, most of the cooking was already out of the way, either in the pantry or in the freezer: I was ready to relax a little and get into the party spirit.

On Christmas Day the men had disappeared for a quick drink down to the Travellers Rest, leaving the girls and I to finish off the lunch. The turkey was coming along nicely and almost cooked before I noticed that the Rayburn wasn't up to its usual temperature. It was fuelled by oil so there shouldn't have been a problem - unless of course the oil ran out. No, I checked the gauge and there was almost half a tank left. I unearthed the instruction book; looking for the trouble shooting section. I rapidly went down the list but, on checking whether the pilot light was still burning, I could see that it wasn't. To restart the pilot light again, all I had to do, so the instruction book told me, was to get a sharp object - something like a pencil - and press the red button, which was to be found inside a little hatch almost at floor level. Collecting a pencil, I scrunched down on the floor, bum in the air, and pressed, as instructed. Nothing. After repeatedly trying this, with no result, I gave up. Now was the time to call in reinforcements. I telephoned the Travellers Rest, asking to speak to Gordon. I could hear the jeers in the background and could well imagine the ribald comments about wives trying to find errant husbands. 'Gordon, could you come home please, I've got a problem with the Rayburn I can't sort out.'

There was a huge wave of noise and then Gordon said, 'Eh? What was that?'

'I've got a problem with the Rayburn, could you come and help sort it out, please, otherwise we're not going to get Christmas dinner this year,' I explained patiently.

'Oh, all right, just let me finish this pint. I'll be no more than ten minutes,' he conceded.

167

An hour later and my cavalry still hadn't appeared over the hill so I gave the Travellers Rest another call. This time I was in an "or else" mood and within ten minutes they had all arrived, beery and stinking to high heaven of cigarette smoke. I explained what had happened and what I had already done to sort the problem out: five sets of hands reached for the instruction book. I left them to it. By this time the Rayburn was all but cold, so I had wrapped the turkey in tinfoil, putting it back in the oven to sit patiently waiting for repairs. I could see that if things weren't fixed - and pretty soon - I was going to have to resort to the microwave.

Despite a bit more poking around with a pencil, the pilot flame remained well and truly out. Luckily the turkey was cooked through, as were the roast potatoes and parsnips; that only left the sprouts, peas, carrots and broccoli, not to mention bread sauce, stuffing etc. The microwave fortuitously performed well, if very slowly, with such vast quantities to cook and re-warm, but at long last we sat down to a fantastic Christmas lunch, which was appreciated by all.

Of course, next day being Boxing Day, there would be no repairs to the Rayburn carried out. Gordon unearthed the camping gas cooker and with this, and the trusty microwave, we managed to cook up many culinary delights. It wasn't until my visitors had all departed, that the Rayburn engineer called. It took him exactly five minutes, from walking in the kitchen door, to fixing the problem. I was stupefied.

'What did you use to press the re-light button with?' the repair man queried.

'A pencil,' I replied.

'Never, never, use a sharp object like a pencil,' he admonished.

'But...but...it says in the instruction book to press it with a sharp object and it actualy mentions using a pencil.'

'Where does it say that?' he queried.

I pointed out the paragraph in the instruction book and he started shaking his head. 'That's just about the worst thing you could do. See...look this is what happened. When you pressed it with the pencil it pushed the button onto the pilot light itself and, of course, it wouldn't start. I've had so many of these call-outs lately, and all for the same reason.'

'I just can't believe it was as simple as that! I've had five men here with their bums in the air looking at it, including a plumber! Oh, well at least it's sorted now.'

It's funny but that Rayburn never was as good as the original one I had in our first barn conversion. Our first Rayburn ran on wood or solid fuel and, when it was alight, was always warm and welcoming. We did occasionally have the odd day when it misbehaved. This usually happened when no one paid any attention to it, or forgot to stock it up. So - sometimes it sulked - but days like that were few and far between and, at least, easily rectifiable by chucking on a bit more wood. This new Rayburn was a state of the art model – all-singing all-dancing. It was highly sophisticated by comparison with its poor little sister, the one that was fuelled by wood. But - a pure wimp of a cooker - easily subdued with a pencil. My old one would have burned it and spat it out in sparks!

CHAPTER FIFTEEN
THE BEGINNING OF THE END OF THE LINE?

We had maximised the full potential of the barn: now came that bittersweet time, when it must be put up for sale. This was always a time in our barn conversions that I disliked the most. We were living in a magnificent ancient barn that was, in essence, a brand new house built inside an old shell. Everything was pristine and new. When you opened the front door you could smell that lovely aroma of wax polish and new carpets. Fabulous. And also very sad. Sad, because very soon we would be starting afresh - shovelling someone else's muck and grime, more than probably living in the meanest, most basic of conditions, all over again.

The estate agent we contacted was extremely impressed with the whole conversion and said that, in his opinion, it would not take many weeks to achieve the asking price. It was definitely about time that we secured our next project. We had already decided that we would like to move to Alderney, in the Channel Islands, where my parents lived, and had been on various agents' mailing lists for some time. We were not going to be able to find a barn to convert, nor were we going to be able to find a building plot, which would have been our next choice of project. Only Alderney born people, and those who had lived on the island for ten years or more without purchasing a house during that time, were allowed to build. Building from new was out for us but at least any EU resident could live there with no restrictions. So...a renovation project was all that would be on offer for us. The only property we had so far found, within our price range, was a house needing extensive renovation, overlooking the harbour. It even had an in-door swimming pool! Gordon thought it deserved an inspection visit so, leaving me behind to dog-sit as usual, he went over to have a closer look.

Gordon was still away in Alderney when the long slow process of eviction for the Grimms started, a good six months after Verity's death. We had all begun to think that the eviction had been cancelled. George and James had long since resumed their daily trip to the shops; after all, it was too much to expect them to change the habit of a lifetime, and we were unsurprised that these trips culminated in their usual drinking sessions. Thankfully, they kept a very low profile when they returned home, and it was now a rare occurrence to hear shouting or any other altercation.

The steady drip drip drip of the eviction process began when I spotted a strange car coming down our lane: it stopped just before the bridge that ran over the stream. A tall woman, dressed very smartly in a dark pinstriped trouser suit, got out of the passenger seat, leaving the driver sat behind the wheel. She walked over the bridge and stood in front of the Grimms' wrought iron side gate. This gate was only a very small gate, probably no more than three feet high, usually kept closed with a padlock and chain. Today was no exception, and she stood rattling it for a while before proceeding further up the lane, towards the only other entrance, which led into the back of the house and the car port. Whether the Grimms were in, or out, this entrance was always barricaded with sheets of steel mesh, old plastic bread trays and other similar paraphernalia. There was no lock of any kind on this obstruction, but to the untrained eye, it looked daunting in its woven complexity. If all this weren't deterrent enough, the pack of mangy curs lurking behind the barrier would have stopped most people in their tracks.

The woman began to shout, 'Mr Luscombe...Mr Luscombe,' trying to make herself heard against the background of frenzied barking. She stood for some time, waiting to see if her call would be answered, before making her way back down the lane to her car.

All this I observed, firstly from the small window in our bedroom, which was set in the side of the house facing the main road, and then the small square window in the front door. The bolt on the gate to the main road sounded very like the noonday gun discharging, when it was drawn, so entrances and exits were very hard to miss. With all the anticipation and speculation regarding the Luscombes' tenuous situation, finding out just who was coming and going, had become imperative, let alone nigh on compulsive, for all of us. It was very nosy I know, but completely understandable in the circumstances.

The smart pinstriped lady came again the next day, pretty well repeating her actions of the day before, with one exception - this time she knocked on our front door.

'Excuse me, sorry to trouble you, but do you know whether Mr George Luscombe is in?' she asked.

'Can you see if there's a white car in the carport? I queried.

'No...no, I don't think there is.' She leaned backwards over the fence to get a better look. 'No, it's definitely empty.'

'Then they must be out. When you see the car in the carport, you can guarantee they're in, even if they're hiding.'

'*Hiding?* Do they do a lot of hiding?' she asked, amused.

'Yes, often,' I volunteered, 'they unfortunately seem to have a lot of people they're trying to avoid lately.'

'I'm probably in that category too then, as I'm from the Bailiff's office and I'm trying to serve a summons on them.' She informed me.

'Ahhah! Yes I'd say they'll definitely be trying to avoid you, if they have the slightest inclination of what you want.'

'Well, they'll have to receive it sooner or later, that's a fact. I'd best try my luck another day rather than hanging around waiting for them,' she concluded.

She could be given top marks for persistence; she just kept coming back. Eventually her doggedness paid off and she finally found the Luscombes at home. After she had spotted the car still in the carport, there was no shifting her, she shouted for George for at least half an hour before being finally rewarded with a response. I couldn't hear the conversation but when George at long last gave in, I saw her hand him an envelope and then leave.

The expected eviction hadn't taken place immediately after Verity's death, giving rise to much gossip and speculation amongst the neighbours, leaving us all even more addicted to the daily goings on of the Luscombes, than usual. Kathleen took to stopping her car in the lane on her way out, or on her way in, on a regular basis, to find out if anything new had happened that day. She would honk her horn outside the front door and hope that one of us would go out to talk to her, to give her all the gossip.

'Hey, guess what I heard,' she chirruped. 'Social Services have been asked to get in touch with the Luscombes to re-house them!'

'There's been a woman here for the last few days trying to get hold of them. She said she was trying to serve a summons from the Bailiff's Office. It must be the eviction notice, I guess,' I said. 'I can't help but feel sorry for them.'

'You must be mad...after all that's happened. I certainly don't feel the least sympathy for them. I shall be glad to see the back of 'em,' replied Kathleen vehemently.

'I know what you mean but they do have a good side, even if they keep it well hidden most of the time!'

'It'll certainly be a bit more peaceful without them around,' she said.

Gordon returned from his Channel Island trip, with excellent tidings. The house he had viewed, although very run down, was in a superb position, with fabulous sea views. He was full of his news and very pleased with himself at securing the property at a good price. I was also pleased with myself, as I had been busy while he had been away with a long string of viewings for our own property.

The day Gordon returned we had another viewing booked. The estate agent was unavailable to show the prospective purchasers around so Gordon did the honours, while I took the dogs for a longer than usual walk, to keep out of the way. I thought I'd given them enough time to view the house in peace, but when I returned I could hear them still talking, above me in our bedroom. I took the dogs into the Atcost building, and then went upstairs to introduce myself. As I walked down the long corridor of the upper hall towards them, the lady in the group greeted me with, 'I'm sorry to tell you but I'm standing in my bedroom! You must be devastated to have to sell such a lovely house but I've just got to have it!' True to her word, we received an offer, through the estate agent, only a few hours later. It seemed as if we were on the move again: time to get the packing cases organised.

A couple of days after this, I was again answering a knock on the front door.

'Hello, I'm trying to get hold of either Mr James Luscombe or Mr George Luscombe. I'm from Social Services. How exactly can I get to the front door to find out if they're in?' she asked in a puzzled tone.

'A difficult one!' I replied.

'You can say that again. I've been up and down this lane a dozen times: it's totally stumped me!'

I trotted out the usual explanations, about the locked gates, the barricades, etc etc, as well as warning her about the dogs. I then asked her, ' Do they know you're coming?'

'Yes, they certainly should know. I wrote and told them I'd be here at eleven and it's past that now,' she said, glancing at her wristwatch.

'It could be that they're trying to avoid you, in that case.'

'You could be right! The word "social services" often puts the fear of God into people, I can't think why. We only want to help them, although I can't do that, without talking to them. I've been told they've got to move out of their home and will need re-housing. All I need to discuss with them, at this moment in time, is their basic requirements.'

'Would you like me to have a word with them, when I next see them?' I offered. 'If they know what you're trying to do for them, then I'm sure they'll want to talk to you. Do you want to leave a telephone number, so they can contact you?'

She thanked me, giving me her card, before leaving.

I didn't see either of the Grimms until next day. James was in his garden doing a bit of weeding. 'James, there was a lady from Social Services trying to get hold of you yesterday.' I told him.

'Oh ah? What did she want?' he growled, looking as if this was the first he'd heard of it and he hadn't a clue what she wanted.

'She said she needed to talk to you, about getting you new accommodation.' I handed him the card with her name and telephone number printed on it. 'That's her number on the card, can you give her a ring?'

'Interfering old biddy. Why can't she just leave us alone,' he grumbled.

'If you're going to have to leave this house, then you're going to need somewhere else to live, aren't you?' I reasoned.

He grunted in derision, 'We aint moving nowhere. If we just sits tight, they'll soon forget about us.'

'That's a bit risky isn't it? What about if you're evicted and have nowhere at all to go? What will you do then?'

'Evicted? Evicted? Don't you myther, maid, we 'aint goin' to get evicted! You just wait and see, us 'ull be here until Doomsday, you mark my words,' he spat.

Talk about being in denial! I still plugged on, though, undaunted. 'Why don't you give Social Services a ring, just see what they want. It can't do any harm to listen, can it? If you find you don't have to move out, then you wont have to take them up on the accommodation they'll be able to offer you. At least that way you'll be hedging your bets.

'Maybees, maybees. Us 'ull see,' was all it seemed I was going to get out of him.

He obviously didn't phone her, as the poor lady turned up again a week later. 'I haven't heard a thing from the Luscombes and still can't get them to come out to talk to me,' she said.

'They're absolutely hopeless. I did ask James to phone you: I tried my best to explain to him, that it would be in his interests to talk to you. I'm afraid I'm not sure I convinced him though. You see - they're determined to stay put. Oh...wait a minute...I think you're in luck, that's them now.' James and George were just coming back through the top gate. By the time they realised that the visitor was for them, not for me, the Social Services lady had got them cornered.

She was a frequent visitor for many weeks but, poor lady, more often than not, failed to talk to them. Even if she did manage to speak to the two of them, over the garden wall, they gave her a bit of a verbal run around, denying that they were going to be evicted. She resorted to knocking on our front door again.

'Sorry to trouble you,' she shrugged 'I really am getting nowhere at all with the Luscombes. All I'm trying to do is tell them that I've found them a lovely little bungalow nearby. They can even take some of their dogs with them if they want to! I told them a couple of days ago about it and that I'd like them to come with me, to take a look at it, but I think they're hiding from me now.'

'Good grief, they're worse than children,' I commiserated. 'They're hard to get through to, aren't they?'

'You're telling me! They've just about worn my patience thin.' She shook her head in exasperation.

'Wait a minute, I think the telephone's still connected.' I'd just had an idea that possibly might work. 'Would you like me to see if I can get them on the phone for you? You never know they just might come out and talk to you if they really are in there.'

'Oh, would you, that would be kind.'

I let the telephone ring for a full minute. We could both hear it loudly from where we were standing, although no one answered.

'Sorry, it looks as if they're either out, or definitely don't want to talk.' I said putting the phone down.

'Huh! Thanks for trying anyway,' she said glumly. 'I had a feeling this might happen and took the precaution of typing out a letter to them before I left the office. Would you do me a great favour and give it to them please?'

'Yes, of course I will. Just as soon as I see them.'

The letter was duly passed on. I did my best to explain to James that the Social Services lady was only trying to help them. He seemed very reluctant to even take the letter from me. It was a difficult conversation. He certainly had it firmly fixed in his mind that they would still be living in the farm until the day he popped his clogs! He didn't realise how much I knew about their situation and, not wanting to cause him undue embarrassment, by giving the game away, I tried to tread as softly as I could.

'The Social Services people are only trying to help you both. I'm sure if you don't like the house they've found for you, they may be able to find you something else. It wont hurt to go with the lady to look at it would it?' I argued.

'I don't know about that, maid. Me and George likes it 'ere. Anyways, we couldn't leave all them beasts, now could we?' he reasoned.

He had a valid point but there were always compromises that might work. 'There may be a small field attached to the house, or you could always try to rent one, if not. Why don't you ask the Social Services lady about that? It should be easy enough to take a few of the animals with you, if you really wanted to.'

'You leave it with me. I'll 'ave a word with our George and us ull see what us can come up with,' James said before making his way back inside the house. *No change there then.*

The Social Services appeared to have given up on their quest for the Grimms and weeks went by before any further developments occurred. The next visitor trying to speak to the Grimms was the RSPCA inspector. The inspector knocked on our door after failing to get the Luscombes to open up. They were more than probably at home, as the car was still in the carport, but they must have been hiding. When they saw the highly visible, sign-written van, it would have left them in no doubt as to the identity of the visitor.

'Good morning. I'm not sure if you can help me. You see I want to have a word with either of the Mr Luscombes,' he said consulting his clipboard. 'Do you know when I might find them at home?'

'Good morning. I haven't seen them at all today: I'm not sure where they're likely to be. Unfortunately it can be quite difficult to get to talk to them, if they don't actually want to talk to you,' I replied. *What an understatement*!

'Well, I really do have to talk to them and sometime soon. We've been asked to call as there's concern about the state of some of their animals.' He was a very solid presence on my doorstep and didn't look the sort of man to give up at the first hurdle.

'I suppose you could wait around, hoping to catch them: you may be wasting your time, though. The best time to get hold of them is early in the morning; say before ten o'clock if you can. That way they may be wandering around the yard feeding the animals, and will more than likely be sober, which is always a bonus.'

He started to chuckle and then said, 'Right you are! Sounds sensible to me. I'll give it a go tomorrow if I get the chance. Thanks a lot.'

He did return the next day but still got no answer from the Grimms. Again I hadn't seen them but did have a sneaky feeling that they were in and watching from one of the many dirt smeared windows. I was out in the garden when he appeared round the corner of the barn and said, with a diffident grin and a raised eyebrow, 'Have you seen either of your neighbours this morning?'

'No, I haven't seen them today: I've no idea whether they're in there or not. In fact, I didn't see them yesterday either, to tell them you were looking for them,' I replied.

'Hmm, not having a lot of luck, am I?' he sighed, 'Do you know how many animals they actually have?' he enquired.

'Now that's a question! Let me see...they've got a flock of probably thirty to forty sheep over in that field there.' I gestured towards the field in the far distance, where we could just see the animals in question. 'And then there's the bull, the cow and calf in the yard: there are some poultry somewhere or other. There're loads of feral cats and I haven't got the foggiest just how many dogs they've got.'

'Are the dogs in the yard the only ones they have?'

'No, there're loads more! There's a West Highland terrier that they keep in the house, plus I think there are some more kept in that deep litter house over there.'

'I can see this is going to be a big job.' he said, ruefully scratching his head. 'It's going to be nigh on impossible to re-house that many working dogs, which does seem a great pity. I shouldn't think that many of them will make suitable pets and I doubt there will be many farmers out there who'll be able to take extra dogs.'

'You've got a bit of a problem, I can see.'

'I'll just have a wander up to the field to see what state those sheep are in, then I'll head back to the office, I think, to report what you've told me. We'll need a fleet of vans for the dogs alone: I'm not sure what will happen to the sheep and cattle.'

'Why are you taking them away? Have the Luscombes asked for that to happen?' I questioned, very nearly sure that they hadn't.

'No, not exactly. We've had more than one complaint lately, about the state of their animals, plus the Social Services Department have also contacted us. I believe they've been asked to re-house the Luscombes in a domestic situation and of course it will be inappropriate to take farm animals with them.'

I could see trouble looming ahead. The Grimms would be fighting tooth-and-nail to keep their animals: who could blame them in the circumstances? The Grimms had always cared for their livestock, at least by their own standards, and were unusually attached to them for farmers. It was going to be a great wrench to part with them.

The Grimms must have sensed that trouble was brewing, as that weekend they asked if we could help them move the sheep down to the Nick Nocks' place again. They said the grass had got a bit sparse but we both had a sneaky feeling this wasn't their only reason. It was entirely conceivable that they had overheard the whole conversation between the RSPCA man and me. Sound carries very well in the country.

We did the usual sheep drive down to the Nick Nocks' and the sheep were safely ensconced in their fresh pasture, on the following Sunday morning, without mishap. They were whisked away just in time - that sheep run was to be our last.

CHAPTER SIXTEEN
DEFINITELY THE END OF THE LINE

The next startling development came, when two large white vans came through the top gate and parked just inside by the grassy knoll. Four very tall muscle-bound young men strode belligerently down the drive. One of them athletically vaulted the little side gate and stood hammering on the front door. I wasn't sure whether the Luscombes were inside or had already left for the day, but the noise of the knocking was enough to wake the dead. Despite the racket, no one came to the door. The other three men had now joined the first one outside the front door. One of them was carrying a sledgehammer, which he smashed with one great heave against the front door. It only took three more of these heavy blows for this seemingly impregnable massive oak and metal-studded front door to admit defeat and cave in. Observing this scene from my usual spy hole in our front door was extraordinary. It was more like a scene from a film set than real life. I was totally appalled; riveted to the spot; dithering about; not sure what to do. In any other circumstances, I would already have been on the telephone to the police. Knowing that an eviction was going to take place sooner or later, left me a little doubtful as to what action I should be taking, if any.

The gang of men went into the house and, thankfully, the Grimms must have been out because they didn't appear with them when they all emerged back into the daylight again. I had, by this time, gathered my wits about me somewhat and gone to find Gordon to ask his opinion. Although we had a feeling their actions must be entirely official, nevertheless, Gordon went to find out if they had any authority to be on the premises. It seemed that they had: they showed Gordon the warrant for possession of the house. It took them no time at all to collect tools from the vans, together with large sheets of plywood, which they screwed and nailed over every downstairs window. The front door was then secured with a substantial padlock before they left. *Well!* What on earth were the Grimms going to do now? The car was still in the carport so they couldn't have been too far away. Perhaps they had been watching from a distance, like me? The men had secured the entrance to the yard, leading to the carport, with a hefty padlock and chain, which they'd wrapped round the mesh used by the Grimms for a barricade along with the stone pillar of the wall abutting it. They were going to have a problem getting the car out that was for sure.

I had to go out myself that day, so the rest of the drama remained for Gordon to observe and relate, when I returned home. Apparently George and James had come round to see him, not long after the men had left, asking if he would come and unscrew the boards attached to the windows. They'd already tried to remove them, but either didn't have the right tools, or were just not strong enough to complete the job. They had decided that Gordon was their best ally. George had said to Gordon, 'Here's a man who knows how to use a screwdriver! Come and give us a hand to unscrew a few boards, boyee.' Unfortunately for him, Gordon had declined to interfere. Gordon told them that he'd seen the warrant for possession of the property and they should stay clear of the house from now on, otherwise they might land themselves and, more importantly him, in hot water.

The advice offered obviously wasn't to their liking. It didn't take them too long to find an old woodwormy ladder, minus one or two rungs, and use it to climb in one of the bedroom windows. Entering through any of these windows wouldn't have been a problem, as most of them had broken panes of glass, but both of the Grimms were in their seventies and, although fit for their ages, wouldn't have been an easy task for them. The ladder wasn't a very long one either, which left James with quite a scramble over the windowsill into the bedroom.

Long after I had returned home, they were still at it. James was inside the house, throwing bits and pieces out of the window to George below. As I drove down the lane, I had seen him throw out a couple of grey pillows. Then came an equally grubby looking eiderdown, which fell on the wet muddy ground, making it even grubbier. George climbed part way up the ladder to take some pots and pans out of James's outstretched hand. They looked like a pair of lovers trying to elope! The indomitable spirit of the Grimms' was hard to fault. Nothing seemed to stop them.

Early next morning I passed the entrance to the carport, on my way for a walk with the dogs. Both George and James were sat on concrete blocks around a small wood fire, which was burning in a desultory fashion on the concrete, just in the entrance to the carport. They were frying something in a blackened frying pan. It looked very much like breakfast alfresco was on the menu at Chez Grimm. The pungent frying aroma, drifting my way, made me feel hungry.

'That smells good, I hope you've saved me some,' I said. The sausages sizzling in the pan were more than a little black but, nonetheless, they still made my mouth water.

'That's good fried bread and sausage that is,' replied George. 'A man needs an 'arty breakfast, this time in the mornin'.

'I'm glad to see you're still in good spirits anyway.' It had rained in the night and, although you could feel that spring was definitely on its way, it was still extremely cold. They were both only dressed in thin shirts, rolled past the elbows, with patched trousers and braces. They looked rough and unshaven but, strangely enough, really no different to normal. In fact with the sharp morning air and warm fire giving their faces a ruddy glow, they looked fit and healthy. The door to the car stood open and I could see pillows and other bedding inside. There was also a very dilapidated sofa on the other side of the car. It had, what appeared to be, straw peeping through the tears and rents but this didn't seem to deter the two cats that were curled up cosily on the filthy cushions. It looked a thoroughly domesticated scene.

'Did you sleep in the car last night?' I queried.

'Yes, James was in the back and I were in the front. It were a bit cramped like, but I slept like a babby. Me legs is a bit stiff now though,' George replied with relish, stretching and bending his legs in their crusty Wellingtons.

'Is there anything I can get you?' I asked, well out of my depth in this situation. It seemed intolerable to me that two old people were living rough in this fashion. A situation I hadn't personally had to deal with before.

'No thanks, nought but the brandy bottle maid! We'em doin' all right.' James took the frying pan off the fire and replaced it with a charred kettle.

'Well, just shout if there is anything,' I said, resuming my walk down the lane and onto the public footpath, with the impatient dogs following at my heels.

There was no doubt about it, this was indeed a beautiful place to live but, I must admit, as far as taking the dogs for a walk was concerned, it was a bit restricting. Apart from walking them on our own land, the only other possible place was a scramble down the public footpath. This path ran along our lane and then through a very rusty five-bar gate just the other side of the Luscombes' carport, past Rick's farm and then on to the village. The gate itself took some moving, as it was below the level of the turf on both sides of it and well dug into the earth. The first time I went through it, I only managed to move it a couple of inches, heaving it aside just enough for us all to squeeze through. After that, I usually climbed over it, letting the dogs scramble through the bottom rails as best they could.

There were two ruts that started as soon as you passed through the gate that were a good six inches deep, almost permanently filled with water, even in the summer, thanks to the heavy clay ground. I presume the ruts must have been made by Rick's tractor, at some time in the far distant past. The foot wide strip between these two ruts was the only place suitable to walk. The rest of the path was profusely overgrown with the wild flowers and grasses, reaching almost to chest height: the disordered hedges attempting to grab me as I walked past, if I should attempt to stray off the central reservation.

On a hot day it was extremely humid, caused mostly by the water that always lay in the tracks, together with the lush vegetation retaining the moisture. With the high humidity, and the loud screeching noise made by the wildlife, I could almost imagine myself in Jurassic Park.

I discovered that it was an excellent place to collect berries in the autumn – blackberries and elderberries –

which together with the apples from our own orchard would make excellent wine. I usually managed to collect a good carton full on each of my treks, which were then stored in the freezer until eventually I had enough to make a batch of wine. When the weather had been dry for a long spell, this was my favourite place to walk.

The solitude of the walk gave me the peace and quiet I needed to contemplate but today, the more I thought about the Grimms' situation, the more depressed I became. Their situation was intolerable: they were obviously living rough in the car and cooking outside on an open fire. There had never been any sign of a toilet outside the house and there was no outhouse that could conceivably contain one. According to Kathleen, there was neither toilet nor hot water inside the house at all. When the Receiver had tried to sell the house, some time before we arrived on the scene, he had instructed an estate agent who had drawn up particulars. Kathleen had managed to get hold of a copy so knew pretty well what was inside. Their ablution facilities had always puzzled us. Where on earth did they go to the toilet? Presumably in the same place they were using now.

On my walk, I debated with myself the logistics of having them to stay with us. It just didn't seem possible to leave them where they were but, equally, I couldn't imagine the scenario of them sleeping in one of our bedrooms. The thought of them taking up residence with us was quite frankly appalling. When I actually put my thoughts into words to Gordon, he was horrified and, I suppose quite rightly, said that they were their own worst enemies. They only had to accept the Social Services accommodation and all would have been well.

Nothing of any importance happened that day. No one visited the farm and the Grimms stayed safely behind their barrier, only occasionally shinning back up the ladder to collect something they'd forgotten. The next day saw the return of Mrs Social Services. The Grimms were sitting targets and had no choice but to talk to her over the barriers. She must have been as appalled as I had been, when she saw the state they were living in. I went round to see them after she had left.

'Are you both all right?' I asked anxiously.

'Yes but that busybody's been back again. I told 'er to leave us alone but she kept on and on at us,' James grumbled.

'She needs to! You can't go on like this, can you? Now you're actually out of the house, you might as well make a clean break and have a nice new bungalow somewhere in the country,' I said.

'Us is all right 'ere. What do us want with a new 'ouse? This 'as allis been our 'ome. We'em very comfy as we are,' said George sullenly.

'It'll be nice and warm for a start. You must be freezing at night now. You can't possibly stay out here forever.'

'Us ull see, us ull see. We've got plans. Don't you worry maid, them gypsies ull get their comeuppance, just you mark my words.'

The old mantra was starting. His alcohol-laced breath was wafting over the wall at me. I might have known they'd not go short on the booze. There was a row of empty bottles sitting along the top of the wall, mainly cider and whisky, should I be in any doubt about the fact.

'Do you fancy a drink, then?' asked George seeing where my gaze lay.

'No thanks, I don't. That stuff isn't going to help you either. It's about time you got yourselves sorted out, instead of just sitting around boozing.' I sounded like a nagging old crone.

'You could get us a couple of bottles when you goes into town next,' wheedled George.

'I might do, if you listen to what that Social Services lady says. Have a bit of sense and think about what she's offering. It might not be so bad after all.' It was fairly evident that I was talking to myself and was not going to get through to either of them in a hurry. 'What sort of cider do you want anyway?' I asked, admitting defeat.

'That cheap stuff from the supermarket'll do. Get us four bottles, maid.'

'OK! OK! See you tomorrow.'

I got the requested cider next day and took it round to them. I hadn't realised that the Social Services lady had just been talking to them; I almost bumped into her as I came out of our gate, with my carrier bag clanking guiltily. She stopped to talk to me, 'I've just been trying to persuade them to visit the bungalow I've got lined up. They're proving a bit intractable to say the least! Is there any possibility that you could try having a word with them?'

'I already have. I've nagged them until I'm blue in the face but I'll give it another go, if you want me to. It's certainly hard to see them living like that but I'm not sure how much notice they take of me,' I replied.

'From what they've said to me in the past, they respect your opinions. It might help with us both digging away at them!'

Well, well, what a turn up. The Grimms actually having a good word to say about me! Some situations do shape up in a surprising way. We said our goodbyes and I wandered off to give the Grimms their supplies. I tried again to reason with them but they were very adept at turning the conversation into other channels and side-stepping the issue. It was obviously a waste of time but I still said my piece, doing my best to paint a picture of them in a new rural idyll.

During this time the animals – all of them, with the exception of the white West Highland terrier, of course – just disappeared. We hadn't even noticed they'd gone until Gordon asked me if I knew what had happened to the cattle.

'I don't know. Aren't they in the far field?' I asked.

'No the field looks empty. Come to that...have you seen any dogs around lately?'

'That's strange: no I haven't. Nor have I heard any sound from the cocks, now you come to mention it. I wonder if the RSPCA have taken them? If they have the Grimms will be as mad as hell.'

We found out later, from Kathleen, that the RSPCA had taken them all away one morning, when we were both out. At least they were all being looked after properly and hopefully re-housed. The RSPCA man must have been a bit mystified about the disappearing sheep.

George and James were still there, living in the carport, many weeks later. The weather had turned very very cold with hard frosts with clear skies. It didn't seem to affect them. In fact the only difference I could detect was that they didn't seem to smell quite as strongly as usual, which was a strange phenomenon. Their washing facilities were even sparser than normal plus they must be hellish cramped sleeping in a car.

James and George had their own particular smells – as of course do we all. When Ted was alive, his odour was the sort you definitely had to take a step back from or, even better, stand upwind of, if at all possible. James and George's aroma on the other hand was more elusive. It was the smell of dead things: a fat white maggotty smell.

Perhaps it was the cold weather keeping the smell down? I don't know. What I did know was that it was still very hard on the nose giving either of them a lift into town. As they hadn't got the use of their own car (it was still parked behind the barriers) they would occasionally ask me for a lift. The windows of the Land Rover would have to be wound totally down, despite the cold, for the whole of the journey and even on the return, when I was on my own, I could still smell that lingering smell of...what exactly? I'm not sure. It totally defied description. They would only ask for a lift on the journey into town returning on the bus, with their carrier bags bulging with shopping, well fortified with whisky from one hostelry or another.

The days the Grimms didn't go into town, would be spent gradually stripping the house of its hoarded treasures. They devised a method whereby James filled a bucket with their bits and pieces, lowering it with a rope to George, waiting under the ladder. He would then carefully lift each item out of the bucket and lay it tenderly on the ground. When the bucket was emptied, it would then be hauled up for a refill. They had already been round to us to borrow a torch as the house now had no electricity connected and, with all of the windows on the ground floor boarded up, they said it was pitch dark inside. Eventually even the torch failed them, and they came round to borrow some batteries.

They soon ran out of places to put their belongings, so came to ask me if they could store some of them in our Atcost barn. I couldn't think of a single reason why not, but told them I would have to check with Gordon first. Gordon, for once, was co-operative and he told them they could store whatever they liked but to bear in mind, that as our house had now been sold, it might not be for long. They assured him that it definitely wouldn't be for long, as they were expecting to get back into the house any day! They deserved top marks for optimism, anyway.

A steady stream of glasses, crockery and ornaments started to appear, packed into all sorts of weird and wonderful old containers, including a couple of ancient rusty biscuit tins. Ted might even have used one of them as a washing bowl!

Most of the items were soaking wet so, one day when I had little to do, I started to help them unpack. There was still a row of scaffolding boards set up on trestles, which had been used as a surface for staining timber, so I set out their wares to dry on this. I also found them a tea chest and some newspaper so that, when the things were dry, they could pack them into it.

There was soon a vast array of goods waiting to be packed away. The barn took on the look of a good car boot sale on a Sunday morning. There must have been about fifty miscellaneous glasses, most of them the sort that had been given free with a bottle of this or that, however, it appeared that one or two had once belonged to a set of fine old delicate glasses. They were probably the same ones we had drunk out of at Christmas; complete with finger marks and old lip smears. There were plates and an immense collection of miss-matched cups and saucers. Lots and lots of ornaments, most of them chipped or cracked. All of these pathetic little items looked like complete rubbish to me but - to the Grimms - priceless treasures with the wealth of memories they held for them.

They took great delight in telling me yarns of how their mother had used a special cup and saucer as her favourite, or which figurine used to sit on Ted's bedroom chest of drawers. Heart breaking stuff really. I did ask them if they wanted to do a car boot sale with some of the less favoured things, as it seemed very unlikely that they would have room for everything in a small bungalow, if and when they ever agreed to accept alternative accommodation. I hadn't been very hopeful when I suggested it but, true to form, they were horrified at the suggestion that they part with even one chipped saucer.

They did want to sell one object, however. This was the granite trough that stood in the front garden. It was of immense proportions and almost immovable but, absolutely beautiful. It had, in all probability, been in that spot since the house was first built. It was worn and rounded with age, delicate algae adhering to the pitted surface. Just why they wanted to sell it was a mystery to me, as if it were mine, I wouldn't have been able to bear to part with it. I think it was only that they couldn't see a way of taking something so heavy with them, which led them to decide to sell it. As our next project was to be on an offshore island, it wasn't an option for us to make them an offer for it, so we asked around in the best place possible - the local pub. It only took one conversation to find a buyer, then next morning Paul, the landlord, and a friend came with a lorry to remove it. It took Gordon, the two Luscombes, Paul, and his mate, to manoeuvre the trough, inch by inch, out of the garden and onto the back of the lorry. Money changed hands and all parties were very pleased with the transaction.

The final tally of the Grimms' repatriated goods was staggering. They had been extremely busy on that ladder. I helped them wrap each item in newspaper then pack it, as economically as possible, into the tea chest, but this was only the tip of the iceberg. Articles just kept appearing; so that we soon ran out of surfaces to put them on to dry out. I found them another two tea chests and these were soon filled too. I had more tea chests but, as we were likely to need them ourselves in the near future, didn't want to part with any more. I thought it very unlikely they would have been emptied and returned to us before we were due to move, in another few weeks time. In the end I visited the supermarket and asked for some empty cardboard boxes for them.

They were stopped in their tracks, however, a couple of days later as someone must have tipped off the Receiver that the Grimms were still around. Either that, or they had been having the house watched, as the four hefty young men returned. They quickly toured the site; searching aggressively in all the obvious places but evidently found no sign of the Grimms. A van and driver stayed at the head of the lane, all that day, keeping a look out. There was, however, no sign of the Grimms. The van left just after dark, returning again at first light. Where the Grimms slept that night was anyone's guess but, wherever it was, must have been extremely unpleasant, as it rained hard for most of it. The van waited again all the next day without a Grimm sighting.

The day after that, a Sunday, the van didn't appear but the Grimms did! We saw them in the lane sawing, with a hacksaw, at the padlock that held the large chain put on by the Bailiff to secure the barrier, which was stopping them getting their car on the road.

'What's going on?' Gordon asked them.

'We've lost that old key to this 'ere padlock and we want to take the car down to Mrs Nick Nock's. They've got a nice dry old barn to put the car in,' replied George. He was looking as disreputable as ever, with his pork-pie hat on the back of his head. His face was as prickly as a hedgehog and he stank to high heaven.

Ignoring his lies, I said, 'I wished you'd just accept that bungalow the lady from the Social Services was trying to give you. It doesn't make sense for you to be living like this, not when you've got the chance of nice warm accommodation.' My exasperation with them must have been evident in my voice.

'We'um going to be near the sheep for a bit. Just to keep an eye on them, like. You never knows whether there be a gypsy or two about, just ready to take one,' said James.

'I wash my hands of you two. You just wont see sense will you?' I said.

'Let's face it; you've lost the house. It's gone. Just resign yourselves to that. All the animals have gone, there's nothing keeping you here. You might as well accept it and make a new life for yourselves,' said Gordon adding his two-pence worth.

'We aint lost nothin' yet, boyee. We'll just see who wins at the end o' the day - you mark my words! We'll be back in that 'ouse before you can blink,' said James vehemently. 'Us ull buy it back if needs be! Us uv got money us 'ave!'

The padlock gave at that point and they started to dismantle the barrier that had trapped the car. There was nothing else that could be said on the subject, sensibly at least, so we left them to it.

On Monday morning as I was taking the dogs for their constitutional, the man sitting in the van at the head of the lane, poked his head out of the window and said, 'Have you seen either of the two Luscombe brothers around lately?'

'Mmmh, can't say I have.' I replied, being economical with the truth.

'Well if you do, you can tell them there's going to be twenty-four hour security from now on. If I catch sight of them they'll get more than they bargained for. I heard in the pub on Friday that they are going to have some help to get back into the house but they certainly 'aint going to do it while I'm around, you can tell them that from me!'

'Help? Who do you mean?' It sounded as if he was under the impression Gordon might be going to help them get back into the house. Was he trying to warn us off? He soon put me right on that score.

'They're calling in a few old debts, so I hear. I also hear that they know a few shady characters but you can tell them from me, I'm not afraid of any of them. HEY...' he shouted. 'Just look over there. There they are! Just behind that hedge.' Sure enough there they were! I could see just the tops of their heads as they walked on the other side of the hedge. When they got to the gap in the hedge, they crouched down, but still could be seen from where we were standing. They obviously hadn't spotted the lookout or, if they had, thought he couldn't see them.

The security guard climbed out of the van and started running towards the field the Grimms were last seen in. I followed him anxiously with my eyes as he caught up with them. Although there was a lot of shouting and arm waving, thankfully that's all it was. I saw the Grimms eventually trudging away over the fields towards Mrs Nick Nock's. They were wearing their greatcoats and both had sacks over their shoulders, bulging with swag. The guard obviously didn't find their stash in the ditch or, if he had seen it, chose to ignore it.

The guard, after that sighting, must have called for reinforcements, as more men arrived a few hours later and pretty soon had the farmhouse looking blind and lifeless. The extra men had brought more sheets of plywood with them, boarding up all the upstairs windows to match the ground floor. The house now was well and truly sealed. The Grimms would have a hard job getting back inside, with those obstacles to stop them. The on-duty security guard mainly sat inside his van, parked at the top of the lane, from where he could only see the front of the house, but he did occasionally make a tour of inspection on foot. With all the windows boarded, the Grimms would need help before they recovered any more of their possessions. Or for that matter returned to the carport to sleep.

Whether the security on the site was too tight for them to risk another visit, I don't know, but that sight of George and James as they tried to hide in the field was to be my last. I was too busy packing the house up for our move to Alderney to go down to see them and regret that I didn't make more of an effort. They had been living in the Nick Nocks' barn for a good three weeks and life was very peaceful, if a little strange, for us without them. As our move was imminent, Gordon loaded up the Land Rover with their packing cases and many boxes, and took them down to the Nick Nocks' barn. They were in the best of spirits, so he told me, and looked as if they had set up home there. They had a fire going, just outside the entrance, with seats constructed from old bricks set up around it. They were still confident that they would be back in their farmhouse before the month was out. They'd even asked Gordon's opinion on how much he thought the house would go for, when it was finally auctioned. They said they had "pots of money" ready and waiting to bid with and were determined to get the house back.

The evening before our move we went to the Travellers Rest to say goodbye to our many good friends. Rumours about the Grimms' doings were rife and, as luck would have it, Mr and Mrs Nick Nock were sat in the restaurant having a meal. After they'd finished, they came over for a chat, bringing us up to date with the happenings in their barn. They had given the Grimms permission to store some of their possessions in the barn, but had not bargained for the fact that the Luscombes would carry this one step further and actually take up residence there! When they found out they were sleeping rough inside the car, parked in their barn, they were horrified and hell-bent on getting rid of them. Not an easy task, as they were to find out to their cost. As soon as they realised what was going on, the Nick Nocks asked the Grimms to find somewhere else to squat as soon as they could. Both of the Luscombes could be very charming, when they wanted to be, and Mrs Nick Nock said they gave her some blarney about only being there for a few days until things were sorted out with their solicitor. They had every intention of buying back the farmhouse. They were only looking after the sheep; keeping the gypsies at bay, etc, etc.

Of course, weeks went by and they were still there. Things came to a head when Mrs Nick Nock went to see them, after they'd had a long lunchtime session in the pub. She should have known better than to tackle them at that point in the proceedings. She must have known, as well as we did, how their tempers could be unpredictable with a dram or two inside them. After arguing the toss for a few minutes, they ended up threatening her with a knife. She managed to back off pretty rapidly: enough was enough for Mrs Nick Nock, she phoned the police. She was pretty upset about the incident - who wouldn't be? They really had done it this time, talk about biting the hand that feeds you! The

Nick Nocks had been just about the only friends they had left.

The Grimms, of course, were well known to the police, who came to interview them but, as far as I could gather, there was very little that could be legally done to move them on. For now the Nick Nocks were stuck with them. I felt sorry for them but couldn't help a niggling little thought that, if they hadn't encouraged the Luscombes to move to their barn in the first place, they might have been snugly ensconced in a bungalow the other side of the valley by now.

We had an early start next day, not helped by the thumping headaches we both had from having consumed far too much alcohol the night before. The furniture van arrived, efficiently loading all our belongings in record time. That left only the clearing up to do, before we could load the dogs into the Land Rover, and leave our lovely home. We were also taking a trailer load of tools and garden furniture, which had been packed a few days before. We'd put our hearts and souls into that barn, not to mention blood sweat and tears, making us very reluctant to leave. But leave we must for pastures new.

Our rig stopped at the top of the drive and I opened the gates for one last time. I stood looking down on the farmhouse with its blind boarded up windows and doors, and gazed across to "our" barn. There were tears in my eyes as Gordon drove through and I closed the gates behind us. A chapter in our lives had also closed with it.

CHAPTER SEVENTEEN
NEWS FROM HOME

The journey to Alderney was truly a feat of planning. Innumerable telephone calls had been made, to work out the best way of getting our furniture, and us, to the island. The only way we could get the furniture, the Land Rover and trailer there was for them to go on the cargo boat that left Weymouth every alternate Thursday. It travelled overnight, usually arriving early next morning but was largely dependent on the weather and tides. The boat normally didn't take passengers but we did eventually manage to persuade the captain to let Gordon travel with the vehicle and trailer. Unfortunately, he wouldn't entertain the idea of me - let alone the dogs - travelling that way. We then had to work out how best to get the dogs and I to our new island home. This turned out to be a passenger hydrofoil, which departed from Torquay, bound for the island of Jersey and then on to Alderney. The three of us were duly booked to sail.

As soon as the furniture van had loaded and left for Weymouth, Gordon took the two dogs and I down to the docks at Torquay, where we were due to catch the hydrofoil to Alderney. Although the tickets had been booked months in advance, the appropriate fare had been paid and even tickets for the two dogs received, as soon as we arrived at check-in there seemed to be some sort of problem. I was asked to step aside to wait while the ticket collector checked with the captain, that it was in order for the two dogs to board. After about half an hour, the message came back that I could board the ferry - but not with the dogs! I had been very close to breaking down when we left the barn, but this was now all too much. Fortunately, Gordon had waited to see us all sail, so I safely left him to sort the mess out while I went for a stomp around the harbour. The captain eventually came out to talk to him and when it was understood that the dogs would be staying on Alderney, not just going for a holiday, he seemed more inclined to accept the booking. What difference it made whether the dogs were holidaymakers or residents was beyond me, but I was, at least temporarily, relieved at his allowing us to travel.

We all eventually climbed aboard and I was told that I must sit on the open afterdeck with the dogs, not in the usual passenger cabin. They provided me with a plastic chair in which I settled myself, shivering slightly in the cold wind, to wait for departure. I was unaware at the time that the shipping forecast was bad for the Channel Island area and due to worsen considerably. We were bound first of all for Jersey. By the time we reached St Helier harbour, I was frozen to the marrow and just about soaked to the skin, despite one of the crew lending me his wet-weather jacket. The spray making its way onto the afterdeck was constant, combined with an extremely keen wind, making it very unpleasant indeed. Quite a few of the passengers, who had been sitting snugly inside the cabin, joined me out on the wet afterdeck - to be sick!

The ocean swells were beginning to look alarming by the time we reached the safe haven of St Helier port, on the island of Jersey. I anxiously asked a member of the crew his opinion on the weather. The forecast was looking menacing but, more crucially, the crewman told me, that as we were travelling on a hydrofoil, which could only cope with waves below six feet, the sea state was a distinct concern too. The waves out in the bay looked threateningly high, with white horses riding on their tops, as far as the eye could see.

The passengers for Jersey disembarked and I could see, looking through the portholes into the passenger cabin, that this had left the ferry pretty well empty. After about ten minutes, one of the crew came to tell me that the boat was going no further and I had to disembark. *Disembark?* This news just about stumped me. How on earth was I going to get to Alderney if I couldn't go on this boat? It was the only boat that went to Alderney from Jersey, therefore, there was literally no other way of getting there, apart from flying.

I had already checked about flights to Alderney, when I was first investigating ways and means of getting us there. The local airline, Aurigny, used small yellow Trislanders with three propellers and only fifteen seats. They were willing to take dogs, at a cost of a child's fare, but each dog had to be accompanied by a handler, which was impossible in our case. Gordon had to travel with the vehicles and now here I was, definitely on my own. If the hydrofoil went no further, then I would be well and truly marooned on Jersey for a week, until the next boat arrived. Which lucky hotel I wondered was going to take me for a week with two dogs?

After pleading with the cabin crewman and explaining my dilemma to him, he went off to find the captain once more. When the crewman returned, he said that the captain had forgotten he'd got dogs on board! The other passengers bound for Alderney had already left the boat, as instructed, but the captain decided that he would try and get me and the two dogs to our destination. By the feel of the swell under the boat at that point, it seemed like a hell-or-high-water situation.

For the journey between the two islands, I was allowed to sit with the two dogs in the usual passenger lounge, in splendid isolation. I was to be extremely glad of this fact, as I am sure I would have been swept off that afterdeck, if we had been still sat there. The waves were horrendous as the ferry crashed its way through, lurching wildly from side to side. Both dogs looked very sorry for themselves but didn't disgrace me by being sick. *Thank God for small mercies.*

When we eventually made the sanctuary of Alderney harbour, there was a welcoming committee of people, all of them waiting to meet the boat. The vast majority of them were waiting for the other Alderney passengers that had been obliged to disembark at Jersey. I was greeted with, 'Where are all the other passengers?' by more than one group of people. I had no idea what had become of them but was just so glad to arrive myself.

The harbour at Alderney has a huge range of tide and, by the time we reached the harbour, the tide was well in. The ferry had two disembarking doors, one directly from the deck (that contained the passenger lounge where I had been sitting) and the other accessed via a very steep, almost vertical, staircase. As the tide was in - my luck was out. We had to disembark from the door on the top deck, by first ascending the staircase. Sounds easy. Not so accompanied by Sharp, an ex-working dog, and now more or less twelve- or thirteen-years-old. I don't think he had never encountered a set of stairs like this one before and he was determined that he wasn't going to do so at his time of life, thank you very much! After the episode along the banks of the Dart, when he jumped into the river rather than climb the metal staircase, he was very wary. Fortunately there was nothing either side of these specific steep steps, otherwise he would have gone over the side, with me still hanging onto the end of his lead!

I had no alternative but to let Tess's lead go, shouting at her to go on up. She was by far the more amenable dog but she was frightened, probably not helped by the vibes emanating from me; she was not about to go up either. I then had to start bellowing at her. Hearing the voice of "she who must be obeyed", Tess started gingerly to inch her way up on her belly, leaving me with no alternative but to pick up Sharp in my arms. He was a Collie and, fortunately for me at that point, had always been on the thin side. Nevertheless, with my rucksack on my back, and Sharp in my arms, it was more like climbing Mount Everest than a staircase on a boat. We did eventually make the top, before my knees buckled beneath me, to find Tess waiting, like the good dog she was, sat at the top of the steps. I got my breath back before emerging onto the deck.

Then came the next obstacle. There was a gangplank joining the boat to the side of the dock. It was made of metal with a rope handrail on one side only, extremely narrow and must have been about twenty feet long from start to finish. It was moving around quite a bit with the surge of the waves hitting against the side of the harbour wall. I knew as soon as I saw it I was going to have trouble. A guardian angel must have been looking on, however, giving me a helping hand in the form of my friendly crewman, who took the rucksack and the end of Tess's lead. He took her across while I staggered behind with Sharp in my arms once again, balancing very precariously on the metal walkway, trying my best to look straight ahead and not down at the heaving seas. God was I thankful to get my feet on dry land! We were all traumatised by our journey and extremely relieved to see my parents who had come to meet the boat.

I was also more than glad to see Gordon, when his freighter pulled into the harbour, early next morning. We didn't wait around long enough for the Land Rover and trailer to be off-loaded; walking up to our new home only a short distance from the harbour, with the dogs glad to have us both back together. After a good look around, we could see that it was a total wreck and would take months of hard slog to lick it into shape. We thought we'd earned ourselves a few days respite, to allow our frayed nerves to settle down, before work began in earnest. Locking the door behind us once more and declaring a holiday, we headed back to my parents house.

It must have been more than six months later before we heard from our old friends Clive and Jayne. A letter came in the post one morning. 'Hey, Gordon, guess who we've got a letter from?'

'Well? Aren't you going to tell me?' he replied

I was too busy reading to answer for a moment. 'It's from Jayne: just wait till you hear all the news.' I passed him the letter so he could read it for himself...

Hello Gordon and Maggie,

Long time no see! How the hell are you both? Hope you're not working too hard! Thought you'd like to hear a bit of news from home. A few weeks ago the Luscombes at last agreed to leave the Stick-Lotts' place and have gone to live in a bungalow in Dunsford village. God alone knows what the locals are going to make of them...they're probably picking a fight in the Royal Oak as we speak! They got mankier and mankier squatting in that barn and just wouldn't shift. The Stick-Lotts said they were going to get a court order to get them out but I think they had some sort of problem with that. Anyway, in the end the persuasive powers of the police and social services combined managed to shift them, by dint of sheer persistence, and they're now living a life of luxury. Oh, I forgot to mention, they were barred from the Travellers Rest! Lots of people complained about the smell and they were so aggressive it got beyond a joke!!

I have been busy with my horses and Clive and Henry have been finishing off the stables that Gordon started. I'm looking forward to it all being finished, so that I can move the horses in, and stop paying rent.

We had a good session at the Travellers Rest last night. Everyone sends their love of course and wants to know when you're coming back. We all miss you!

Lots of love from all of us

Jayne, Clive and Keeley
xxxx

Printed in the United Kingdom
by Lightning Source UK Ltd.
115708UKS00001B/33